Keeping Chickens

Garden Farming Series

Keeping Chickens

John Walters and
Michael Parker

PELHAM BOOKS

First published in Great Britain by
Pelham Books Ltd
44 Bedford Square
London WC1B 3DU
MARCH 1976
SECOND IMPRESSION JUNE 1976
THIRD IMPRESSION FEBRUARY 1977
FOURTH IMPRESSION NOVEMBER 1979

ISBN 0 7207 0882 6

Printed in Great Britain by
Hollen Street Press Ltd, at Slough
And bound by James Burn Ltd
at Esher, Surrey

To
Jan, Jane and Henrietta

Contents

Conversion Factors

Throughout the text Imperial units (e.g. feet, inches, pounds, ounces, °F) have been used for measurements of houses, runs, feeds, temperatures, etc., since most of us are still a long way from 'thinking metric'. For information a useful table of conversion factors is given below.

Imperial Units to Metric Units

1 in.	=	2.54 cm
1 ft	=	0.3048 m
1 mile	=	1.6093 km
1 gal	=	4.5461 litre
1 sq. in.	=	6.4516 sq. cm
1 sq. ft	=	0.0929 sq. m
1 sq. yd	=	0.836 sq. m
1 acre	=	0.4046 hectare
1 oz	=	28.34 g
1 lb	=	0.4536 kg
1 ton	=	1.016 tonnes
$x°$ Fahrenheit	=	$\frac{5}{9}(x-32)°$ Centigrade

Metric Units to Imperial Units

1 mm	=	0.0394 in.
1 cm	=	0.3944 in.
1 m	=	3.2808 ft
1 km	=	0.6214 miles
1 litre	=	0.2199 gal
1 sq. cm	=	0.1550 sq. in.
1 sq. m	=	10.764 sq. ft
1 sq. m	=	1.195 sq. yd
1 hectare	=	2.471 acres
1 g	=	0.035 oz
1 kg	=	2.2046 lb
1 tonne	=	0.9842 tons
$y°$ Centigrade	=	$(\frac{9}{5}y+32)°$ Fahrenheit

Why Keep Chickens?

Laying hens are returning to back gardens—not on the scale of pre-war and immediate post-war days, when every other garden seemed to boast a flock of six or twelve hens, but in sizable numbers if the chick breeders and manufacturers of housing and equipment are to be believed.

Reasons for the return are varied and range from fears that the price of shop eggs will go through the roof to a feeling that a back garden flock is the only way to get really fresh eggs. There is also the desire for a profitable hobby which will break the hypnotic hold of the TV screen.

No one would dispute the freshness of the eggs coming from a domestic flock, none could be fresher, and your chickens will certainly keep you away from the television—they are a 365-day-a-year job. They will have to be fed, watered and managed every day and if you are unable to be there, you will have to find a co-operative neighbour to do it for you.

It is a job which has to be done every day of the year, regardless of the weather and regardless of the disposition of your liver. If the birds are going to survive, flourish and provide for the breakfast table, troughs and feeders will need replenishing and the drinking points will have to be topped up. A close eye will have to be kept on their general condition. You will need to recognise immediately any sign that they are below par and take appropriate action.

In time, you will develop that 'sixth' sense which no one can teach but which comes with experience; it is referred to as 'stockmanship'. This ability will enable you to nip trouble in the bud before it occurs and to provide just the type of environment which will get the best out of your flock.

Stockmanship flourishes best in a small flock. It tends to get lost in a vast number of birds where stockmen are not even able to see the hens well enough to clear out the dead ones, let alone look for signs of poor health in individuals. In the commercial world, the brown-egg layer averages about 240 eggs a year. In the back garden, small flocks of certainly under fifty and more often near twelve or eighteen birds have regularly topped 300 eggs a bird a year. This is not so far removed from the magical 100 per cent egg-a-day target.

But do not be under any illusions—considerable effort must go into producing the results, both in time and money, because it means feeding the best rations available. It is not in terms of hours a day—total time will not be as much as 60 minutes a day, once a routine has been established—but it is a fact that has to be faced. The work has to be done every day regardless.

Having put the dark side of the job, let us look on the brighter side which should far outshine the problems. First there is that regular supply of eggs—all the year round so long as you are able to light your flock in the winter (see Chapter 5).

Second, the confidence that the eggs are of top quality and just as fresh as you want them to be. The quality of shop eggs is, on the whole, reasonable, but often no more than that. With your own flock and the knowledge that will come with experience, you will be able to manage so that the quality is the best possible.

Egg farmers have not yet developed a system of telling the customer how fresh the eggs are that he is buying. Under Common Market regulations, shop eggs have to carry on the packet the number of the week in which they were packed, but not laid. If they are sold from Keyes trays there should be a display card giving the same information. Even then you need to be something of an expert to decipher the code markings and know the number of the current packing week.

The third advantage is that of a creative, satisfying and economic hobby which will save money so long as your poultry farming activities are regarded as a hobby. If you start to charge your labour at even 60p an hour it ceases to be economic, but one assumes that you do not attach a monetary value to your leisure hours.

There is no reason why your birds should not produce eggs as economically as the commercial producer. He buys his feed and birds at a better rate because he gets the advantage of the economies of size. His feed comes by the ton, you buy by the $\frac{1}{2}$ cwt. His birds come by the thousand, you buy a dozen or so at a time.

It is dangerous to quote prices because they are always on the move, but on average the domestic poultry keeper pays about £5 to £6 a ton more for his feed than the commercial man, an extra 30p to 40p for his point of lay pullet and £1 to £2 a bird more for his housing and run.

His production costs, calculated on four to five eggs a bird a week, are in the region of 37p to 39p a dozen. This takes account of feed, cost of bird and puts a value of 50p on each carcase at the end of the laying year. It can be even cheaper if the birds are moulted successfully and taken on for a second year, (see Chapter 5).

This 37p to 39p, however, takes not account of the cost of housing and equipment, such as feeders and drinkers, nor of labour. Marketing is another cost with which the commercial industry is saddled and which the consumer has to pay for in the end. The packing station has to be paid for grading and candling the eggs. Then there is the wholesaler and the retailers who all have to take their cut. This is why there is a difference of 20p to 25p a dozen between what the producer is paid and what the housewife pays in the shop.

By producing his own, the domestic poultry keeper is saving this distribution margin which, as transport and labour charges continue to rise, will increase. At the moment, the price of the egg nearly doubles on its way to the housewife.

In an age when the consumer is becoming increasingly remote from the source of production of his food and reliant on the pre-packaged, processed, frozen and sometimes flavourless goods on the local supermarket shelves, do-it-yourself egg farming would seem to fulfil a deep psychological need for a natural product untouched, as it were, by other human hands.

But a final word before we get stuck into the instruction, and before you set up the house and a run and buy-in the first birds. Check with your local council to see that there are no local by-laws which prevent you from keeping chickens in your back garden. Some councils do have their own thoughts on the practice and they may lay down regulations as to how many birds and how far they should be from any dwelling place. So better be safe than sorry and find out first what the position is in your part of the country.

Another source with which to check would be your neighbours, if you have any. It seems only good manners to let them know what you are planning. The chances are that if they wake up one morning to find the garden next door full of chickens they will object on principle. If they are going to overlook your poultry house it is also incumbent on you to choose a design that is aesthetically pleasing.

It is also up to you see that the unit does not give off any offensive smells and does not attract vermin. A certain aroma is inevitable, but good husbandry will keep it from becoming objectionable. Vermin will only be attracted if you are careless and leave food lying around. So clear up after the feeding rounds and make sure everyone else in the family follows suit. Food is costly enough without scattering it around to attract rats and mice. There is no profit in keeping vermin fat and your neighbours will appreciate the gesture.

Good luck and may the breakfast egg that comes direct from your back-garden flock be one of the highspots of the day.

2 Housing and Equipment

Over the years, commercial egg producers have for a variety of reasons adopted the battery system in which to house their flocks. Cages not only enable large numbers to be kept in a relatively small area, but they can be serviced easily and relatively cheaply—a vital consideration for the cost-conscious farmer.

Even within the cage system there are many variations: tiered, with the cages directly above each other in three or even four levels; Californian, where the levels are in an 'A' shape so that the manure can fall directly into a droppings pit; and flat deck, where all the cages are on one level.

But whatever the system, everything is geared to ease the load of the poultry worker. The birds can be fed at the press of a button. Their manure, if it does not fall straight into a pit which can be cleaned once a year, is also cleared automatically.

Even the eggs can be collected by pushing a button and many of the large units holding 100 000 birds

Fig. 1 This house is large enough to take up to 45 layers. For 15 birds, just the first third of the house is needed, or the first two thirds if the flock is 30 strong. The three batteries of nest boxes (right) each contain four nests and are placed 5 in. or 6 in. above the litter. They face away from the source of light so will be dark enough to encourage the birds to use them for laying. A droppings pit system is used instead of perches and droppings board. The slats should be removable to give access for cleaning out the droppings. This is a job which can be done every six months or so, but it can be left until the birds are removed. Drinkers can be suspended above the slats at a convenient height for the birds who will roost there at night.

upwards work this way. Webbing belts run along the front of the cages and when the right button is pressed the eggs are conveyed to the end of the cage block where more inspired gadgetry carries them up or down to one level.

Instead of traipsing up and down the cage row, the egg collector takes the eggs off the belts and places them in papier mâché trays (thirty eggs to a tray)

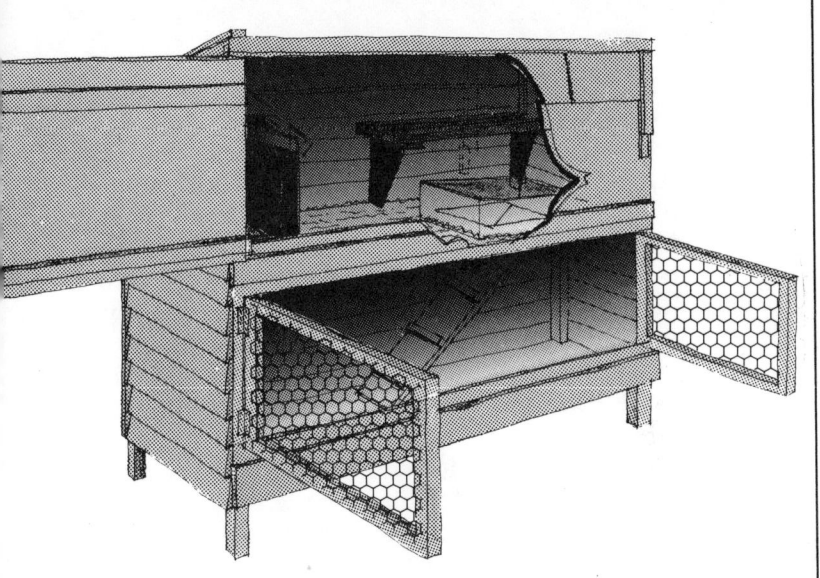

Fig. 2 The latest housing idea for just three or four birds is this two-storey house. The nest box, litter area and perch are on the first floor and the open 'range' on the ground floor. A wooden walkway acts as a staircase between the two levels. Known as the Selsey, the unit has been designed by Southern Pullet Rearers, Chichester, in conjunction with Lewes Road Sawmills Ltd. See Appendix A for full addresses.

which are then taken to the packing station. Here they are inspected against strong lights for blood spots, cracks and other faults, graded according to weight, packed in cartons of six or twelve and whisked off to the supermarket. It is modern times with a vengeance and as efficient as it needs to be for a modern, up-to-date industry.

An egg farmer can pay up to £6 or £7 a bird to cage and house his flock, and a unit of 100 000 birds will not leave him much change out of half a million pounds. It is a far cry from the humble domestic poultry keeper with a couple of dozen birds in the back garden, but it helps to put in perspective the price of eggs in the shops and to explain why modern egg production is being carried out by larger but fewer units.

Fortunately, the chap in the backyard does not have to measure his housing costs in thousands of pounds, but he will have to think in terms of at least £35 or £40 if he is going to buy a ready-made building to hold six to eight laying birds and provide a run for their use during the day. The British poultry industry was founded on hundreds and thousands of such sheds measuring about 6 ft × 4 ft and just the right size to take a handful of layers or a small breeding pen of show birds.

Put a partition down the middle and it could be used for a couple of pedigree cockerels; put a coop in the centre and it provided comfort for a broody hen and her family. Whatever its use, the old 'six by four' became the foundation stone of the modern industry and it is still in use today.

Basic requirements for a house for laying stock are: a floor covered in shavings or peat moss; two or three perches across the back; and about 10 in. below, a board to take the droppings. Birds excrete half their droppings while they are on a perch—it is easier for them—and removal can be a simple matter of scraping the board about once a week.

The droppings can be used on the garden, but they may need to be dried. If this is done as they are removed from the house it should be no problem, although you may get a backlog of manure during a spell of wet weather. The droppings make a splendid all-purpose fertilizer, particularly good for grass, but do not apply them in a fresh state. It tends to burn the vegetation.

Poultry droppings are lacking in potash but this can be remedied if you mix them with fresh bonfire ash. The best time to use them is in the spring and they make a splendid activator for the compost heap.

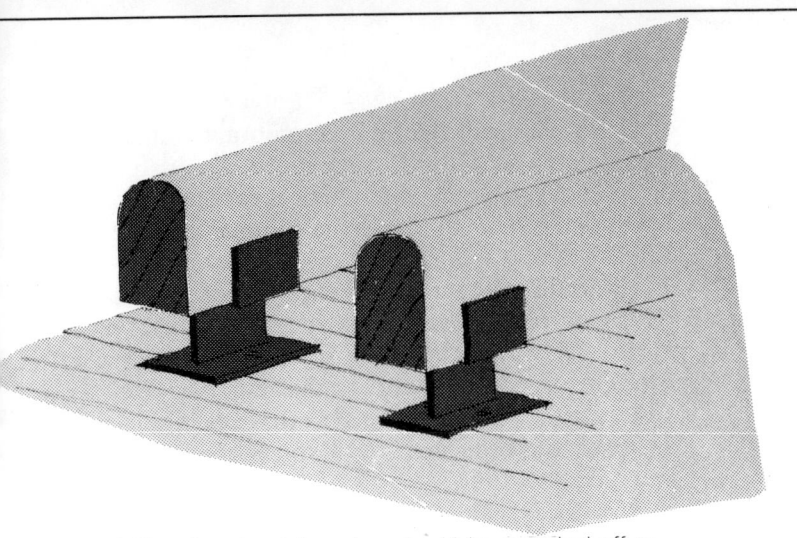

Fig. 3 The top edges of perches should be smoothed off to fit the birds' feet. Make the perches removable so that the droppings board may be cleaned easily.

Set the droppings board 2 ft 6 in. above floor level and make it about 18 in. wide. Perches should be set at the same level as each other and not less than 15 in. apart. Hens are funny creatures and if one perch is higher than the others this will be the only one used. They do not like to be looked down upon and if nest-box tops or any other vantage points are higher than the perches they will become the favourite roosting places.

New arrivals in a house often have to be trained to use the perches. They will roost anywhere but the place you intended. If this happens, you will have to call back after they have been shut up for the night and place them on the perch yourself. After two or three visits they should get used to the idea and perch without assistance.

Make the perches removable. It will save a lot of bending into awkward places at cleaning-out time. An alternative is to fix them to a metal bar which can be swung out of the way to leave the droppings board clear for cleaning. Round perches are not the most comfortable so avoid broom sticks. Go instead for a piece of 2 in. × 2 in. with the top edges smoothed down. It is a size that best fits the feet of adult stock. See Fig. 3.

Minimum size for an individual nest box is 1 sq. ft. It can be either free-standing in a block of two or three, or fixed against the wall. It should be lined with enough litter to make it comfortable. A mesh floor underneath will improve air circulation. Placing of the boxes can be vital. They should be in a dark corner away from the light—birds do not like to perform the very private function of laying an egg in the full light of day. A suitable nest-box is shown in Fig. 4.

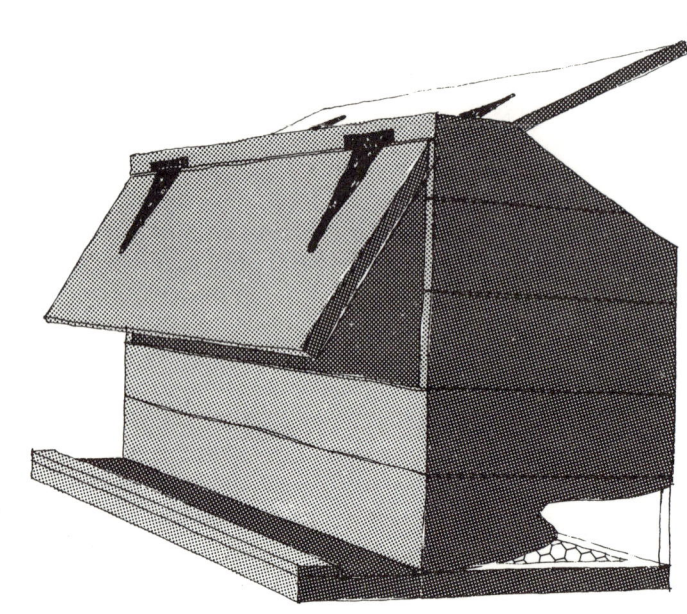

Fig. 4 Nest box with two lids. The top lid gives the egg collector access to the nest without the need to enter the house. The front lid can be held open on a cord during the day and let down at night to prevent the birds sleeping in the nest. Note the alighting perch and the wire mesh base to improve air circulation. A flow of air will keep the nesting material in good condition.

Take care to get the level of the boxes right. Set them too high and the level of the roof will conflict with that of the perches, although a sloping roof will prevent birds from perching. If the nests are too low on the floor the eggs will run the danger of being eaten by the birds and hens will not get the right degree of privacy.

Raise the boxes at least 5 in. above the floor and the hens will soon learn to jump up and settle in the dark, comfortable seclusion. An alighting rail along the front will provide a convenient landing point.

They may find the boxes too comfortable and try to use them as sleeping quarters. A bar across the entrance will put a stop to this trick. It can be dropped in place last thing at night and removed in the morning.

Having got it right for the hen, how about the egg collector? What is the best way to ease his labours?

One of the best arrangements is to fit the boxes on the outside of the house so that the egg harvest can be gathered without disturbing the birds. Just lift a flap and there are the eggs. In cages, the egg simply rolls down the sloping floor out to the curved shelf at the front to await collection.

The same system of semi-detached boxes can be used for the food trough and the drinkers. The birds are able to get at the feed and water through grilles on the inside of the house, and these can be topped up from the outside. A lid will keep the feed dry and prevent the water from freezing in the winter. Alternatively there are a number of proprietary brands of feeders and drinkers on the market.

For wet mash there is the V-shaped trough with a spinner over the top to keep the birds from treading in the feed. This feature is particularly important in any economy drive. Too much space at the top of the trough and the birds will be in there treading and

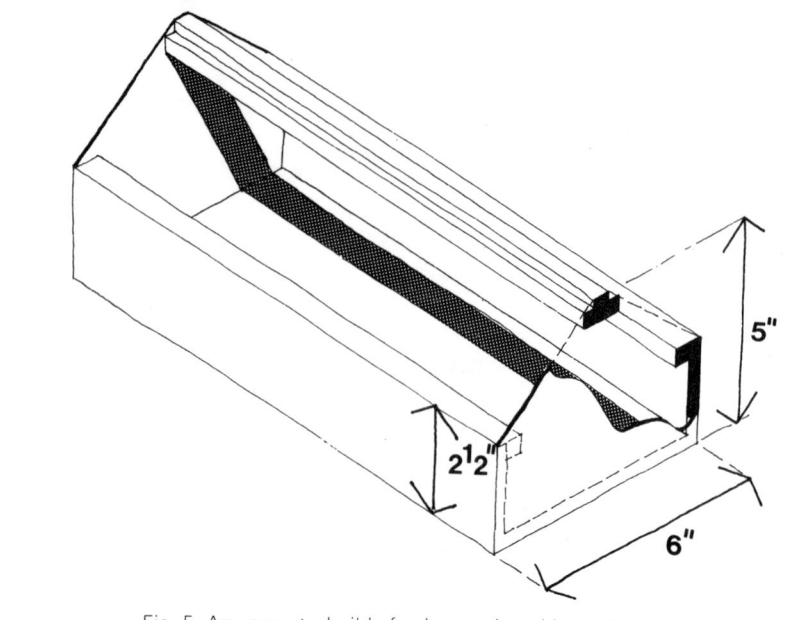

Fig. 5 An easy-to-build feed trough with anti-waste lips running along both sides and across the ends. The lath on the top should swivel to stop birds perching and will keep them out of the mash. It also serves as a carrying handle.

scratching, creating a carpet of expensive feed which will be trodden in and wasted.

Troughs can be bought ready-made, but it is not difficult to build your own out of odd pieces of wood. Fig. 5 shows a suitable design. The lips that run the length of the trough on either side will prevent waste and the top lath will keep the birds out of the mash. It also serves as a handle. Attach the top lath so that it swivels at the slightest touch. This will deter birds from perching over the mash and fouling it with their droppings.

Obviously the trough should be deep enough to hold sufficient food and be long enough to allow all the birds to eat at once. Too little trough space, and the birds lowest in the peck order will not get a look in at feeding time. This means an allowance of about

Fig. 6 Galvanised tube feeders will serve 25 birds. The feed
goes in at the top and flows out through the grille at the base.
It can be suspended from the roof and should be at a convenient height—about one inch above the level of the bird's
back.

4–5 in. of trough space for each bird, remembering
that most troughs are double-sided, unless they are
the hopper type set in the wall. So if you have twelve
birds you will need 48–60 in. of trough space, but
not necessarily all on one trough. Better to have a
number of easily-movable small ones than one long,
cumbersome piece of equipment.

For dry mash there are the galvanised tube feeders.
These are basically cylinders which are topped up so
that the feed flows out into a pan at the foot. They are

built so that clogging inside the tube and waste are cut to a minimum and they really are the best bet for the beginner. A hook or handle at the top enables the feeder to be suspended from the roof. They should be suspended about 9 in. off the litter and there should be one tube feeder to every twenty-five birds. (See Fig. 6.)

With costs rising all the time, it is difficult to be precise on price, but feeders start at about £2.50 and come in a variety of sizes.

There is a similar arrangement for poultry drinkers with a central reservoir feeding a trough at the bottom. The drinkers should be topped up at least once a day and it is vital to buy one that has sufficient capacity for this operation. As a rule of thumb guide, each bird needs at least half a pint of water a day. Water is cheap but vital to egg production, so be generous and work on a daily basis of half a gallon to six birds, one gallon for twelve birds, and so on.

One more allowance to remember is floor space. Each bird needs about 3 sq. ft of floor space in which to move around, so if you are thinking in terms of a 6 ft × 4 ft building keep your flock size down to eight birds, or ten if they have access to range or to a run which they can reach through a pop hole or hatch.

Before moving outside the house, however, a few words about the base which can be either earth or wooden. If you decide on wood remember that most shed manufacturers quote a price excluding a floor. That costs extra.

A floor gives you the advantage of being able to raise the house off the ground on concrete or brick piers. There is always a certain amount of dampness in the ground which will rot wooden footings in time and a gap underneath will deter rats and other rodents. It is much easier for a dog or cat to seek out rats under a house on stilts.

In some cases it is possible to fit the piers with a rat baffle to at least keep the rodents out of the house. A baffle is simply a cone with point uppermost which fits like a collar round each pier.

If you still prefer an earth floor, ensure that the soil is rammed down hard before the building is erected, so that it forms a solid base on which litter can be spread.

Litter material can be soft wood-shavings, peat moss or straw to a depth of 6–9 in. Rake it over regularly and break up any lumps which become matted. If the droppings board and perches are doing their job the litter should remain in good condition for some time.

Obviously the size of the garden will, to a large extent, govern the size of your run. Ideally the birds would look best with a grass run, but unless the garden stretches to at least half an acre, forget the idea. Any smaller area will soon lose its grass and become an untidy mud patch and a disease risk to the birds.

If you have less than half an acre to play with, go instead for a run which covers about twice the area of the house. If the house is 24 sq. ft the run should be 48 sq. ft, for a 36 sq. ft house the run should be 72 sq. ft and so on. It is as simple as that. The only difficulty is finding the space.

There are two ways to avoid the mud heap already mentioned—the birds can either parade on a scratch area of cinders and clinker, or they can be raised up a couple of feet from the ground on a verandah.

For the scratch area, dig out the area of the run to a depth of about 6 in. and pack it tightly with hardcore, topped off with cinders and clinker. Such a surface will drain well and be quite acceptable to the birds. Do not site the run in a hollow into which rainwater will drain.

One amenity your flock will appreciate is a dust bath. As well as an antidote to boredom it will help keep them free from insects. Fine dry earth, sand or coal ashes placed in a shallow box will make them your friend for life. If the run is covered the bath can be placed in the corner. If there is no roof to the run it will be necessary to provide a small shelter just for the bath.

Ideally the area of the run should be covered by an extension of the house roof, but it should certainly be surrounded by wire netting which will withstand attacks from your dog, next door's cat and any passing foxes. The roof will keep out the worst of the weather as will a board round the foot of the run. A fence, a hedge or some sort of windbreak close to the run will be appreciated by the birds when there is a strong north-easterly blowing.

A 2 ft high verandah can be built of wooden slats or weld mesh. Slats should be about 1 in. wide and spaced about 1 in. apart. The mesh should be 3 in. × 1 in., 10–12 gauge wire.

Cages are another possibility for the back garden, particularly where space is limited. They will fit into a shed or a garage and have the same benefits for the domestic poultry keeper as for the commercial man. With one, two or three birds to a cage, it is easier to see if one goes off lay. Individual records of production can be kept and ailing birds can be spotted immediately. Droppings can also be watched for any signs of ill-health.

There are a number of types of cage on the market, usually available in units of six birds. They are equipped with feed and water troughs although the more sophisticated have nipple drinkers (see Fig. 7), much favoured in the commercial industry. A header tank holds the water which is carried by plastic tubing to the nipples suspended one between every two

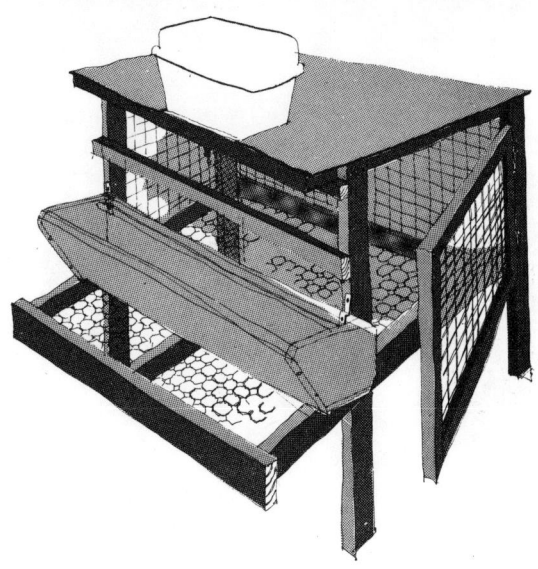

Fig. 7 This latest line in domestic battery cages is called a Home Egg Unit. It will fit into a shed or garage and holds six birds in the two cages. The header tank feeds water to nipples at the front of the cages.

cages. The bird's beak activates the nipple and it soon learns to operate the valve which releases a supply of water.

Fold units and range arks (see Figs. 8 and 9) offer an alternative method of housing to anyone with more space than is provided by the average garden. They are lower in height than the normal shed and are portable, in that they can be moved frequently to a fresh patch of ground. Slats are used for the floor and the droppings fall through, fertilizing the ground as a by-product. Feed and water points are fitted outside the run for easy filling.

Labour costs have ruled out this system in commercial practice, but it can provide a cheap form of housing if the land is available. Wheels will obviously help as will relatively level ground, but to be successful the fold must be kept on the move.

Fig. 8 The range ark—designed to be moved about over a stretch of open range so that the birds get a fresh patch of ground every few days. This ark measures 6 ft × 3 ft and with two nest boxes taking four birds to a box it will accommodate eight hens. By increasing the number of nest boxes the number of hens can be increased accordingly. The ark can also be used for rearing when it will take up to twenty pullets depending on the size of the range.

Fig. 9 The fold unit. Like the range ark, this unit can be moved to fresh ground as required. With four nest boxes in the fold unit, up to sixteen birds can be housed. The size of the unit shown is 11 ft × 6 ft including the run.

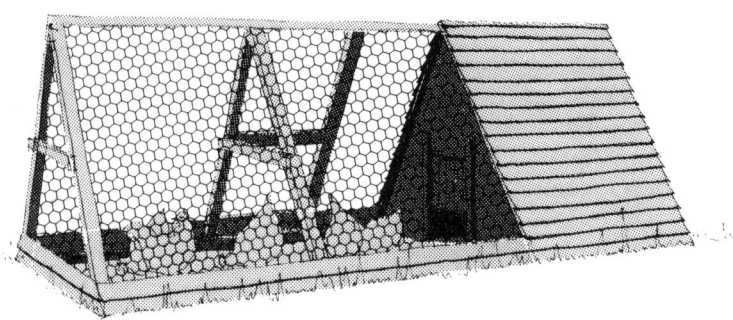

Which Bird?

The basic job of a backyard flock is to convert feed into eggs and to do this economically it is vital to choose the best type of bird for the job. You want a strain that will produce the type of eggs you want on the smallest amount of feed. It must also be able to live and thrive in the conditions you provide and be a good converter—consuming just enough to keep body and soul together and maintain production during the laying cycle.

The commercial poultry industry has this down to a fine art and over the years has developed layers which will happily average 240–260 eggs a year on a relatively miserly 4–4.5 oz of food a day. These birds, of course, are kept in intensive battery conditions, but there is no reason why the backyard poultry keeper should not approach and even beat these figures with good management and the right type of bird.

It is probably easier for the novice to buy his flock as point-of-lay pullets. This means an eighteen- to twenty-week-old bird that has been reared from day-old by a specialist pullet grower.

Obviously they will be more costly than a day-old chick—somewhere in the region of £2 to £2.50 a bird—but the novice will not have had the expense of feeding, mortality (a few are bound to die in the early stages) or vaccination against fowl pest, infectious bronchitis and Marek's disease (see Chapter 6 for details of these diseases). He will also be saved the expense of putting up and equipping a separate house in which to rear the young stock. In a small garden separate rearing quarters are not always practicable.

Don't be tempted to buy-in birds that have already started to lay. The idea of instant eggs is very tempting, but it is difficult to make the move to new quarters without putting the birds off lay and this condition could last for some time—days, if you are lucky, weeks, if you are not.

A three or four week settling-in period before the birds get down to the serious business of egg production at around twenty-four weeks of age enables them to adjust to their new quarters and you to get to know them.

Just a word on the difference between a pullet and a hen. A bird remains a pullet until it has completed its first laying season. Spring-hatched chicks come into lay in the autumn and assuming they are lit during winter (see Chapter 5) will continue to produce eggs until the following autumn.

At this point the birds will start to go into a moult. Feathers will start to fall out all over the place and new plumage will start to grow. Strictly speaking, this is when the pullet becomes a hen, but in practice the term 'pullet' is applied to the initial weeks of the first year of lay.

The commercial industry deals, with the occasional exception, with first year layers. At the end of fifty-two or fifty-six weeks in lay, most birds are carted off to the poultry packing station and point-of-lay stock takes over in the cages.

But in the back garden it is not always necessary to be so ruthless. If some members of the flock are still performing well there is no need for them to be culled at the end of the year. In the interests of economy, it might be best to retain, say, half the flock and buy-in replacements. The eggs of the newcomers will help fill the gap created as the old hands go through the moult and egg output dries up for ten to twelve weeks. Management of the moulting bird is discussed in Chapter 5.

Obviously if you are going to cull the poor doers in a flock, it is vital to keep a close check on each bird's individual performance. This is a simple matter with caged flocks. You know immediately those that are not laying. It is even possible to keep individual production records for each member of the flock. But it is not so simple with birds on the floor and here a better guide is the general condition of the bird—bright of eye and of comb and a general air of alertness and well-being. (Other hallmarks of a good layer are described in Chapter 5.)

But once you have decided to cull, remove the birds before they reach the moulting stage. Again it is a question of economics. You do not want a bird eating its way through expensive meals if it is not going to live long enough for you to reap the benefit.

There was a time when there were hundreds of types of different breeds of chicken about, but many of the more exotic types have disappeared with the growth of the modern industry. It is the day of the hybrid—a commercial bird developed by the major breeding companies of the world. They have taken the best features of existing breeds and by a long process of breeding by selection have brought them together in one strain. This is then given a name or even just a code number and marketed by the million throughout the world.

The popular breeds are still available, but not so readily as they were fifteen to twenty years ago. It is easier to spot traces of the Rhode Island Red, the Light Sussex, White Leghorns and Wyandottes in the code-numbered hybrids of the international breeding companies.

Assuming you want a bird for egg-laying capabilities and not for its ability to put on meat at a relatively fast rate, the first decision to make is what colour egg—white or brown? Having settled on the colour, there is a question of size of bird. Do you want one of the heavy breeds or hybrid strains which eat perhaps an extra half ounce of food a day, but produce a good size carcase which will provide the family with a slap-up Sunday lunch when the time comes to knock her on the head?

If you are not too interested in your flock as table birds, or if the family could not stand the thought of sitting down to a lunch of Henrietta who has served them so faithfully during the year, go for one of the lighter types. Here, as you would expect, appetites are smaller and feeding not so costly.

A halfway house between the two extremes is a light-heavy cross where the laying characteristics of one breed are combined with the meat-forming qualities of another. It is the sort of operation the professional breeders are carrying out all the time on a more sophisticated scale. Their multiway crosses are worked out by a computer.

The Rhode Island Red has always been a popular bird for crossing—RIR × Light Sussex, White Leghorn × RIR, and White Leghorn × Light Sussex are well used combinations for egg production. Left to their own devices pure breeds can be slow to mature and only reasonable egg producers. So long as the genetic make-up is complementary, cross breeding seems to act like a tonic and speed up the process, leading to faster growth rates, fewer deaths and more eggs. They call it 'hybrid vigour'.

But back to the pure breeds and the main differences between the light and heavy varieties. Apart from the obvious fact that they are larger (weighing 5 lb to 7 lb), the heavies tend to lay brown or tinted eggs and are more prone to broodiness (see Chapter 5) than the lights. Weight of the lighter birds is between $3\frac{1}{2}$ lb and $5\frac{1}{2}$ lb, the eggs are white and the feathers grow more quickly than with the heavies.

The Rhode Island Red is the best known of the heavies. It can be a prolific layer of deep brown eggs, but unfortunately its table qualities are not the best. The Light Sussex is a better combination of egg laying and meat qualities. Its eggs are light brown and it weighs between $5\frac{1}{2}$ lb and $6\frac{1}{2}$ lb. Plumage is basically white with a ruff of black feathers round the neck and a black tail. Another popular heavyweight is the chunky white Wyandotte which originated in the USA and also lays light brown or tinted eggs.

Among the light breeds there is the White Leghorn which is apparent in so many of the commercial hybrids. The major breeders like its ability to lay plenty of white eggs on a relatively small appetite. It weighs in the region of $4\frac{1}{2}$ lb. The Ancona, like the Leghorn, comes from Italy and lays a lot of eggs over a long period. Feathering is mottled black with a flowing tail.

These few names represent the tip of the iceberg. There are hundreds more—or perhaps it is correct to put it in the past tense—there *were* hundreds more. You may have to search for the Rhode Island Red, Light Sussex, Welsummer and Wyandotte. Try your *Yellow Pages* or the classified advertisement columns of your local newspaper. You will find the hybrids advertised, but occasionally the pure breeds will be there too.

Failing that, the Poultry Club of Great Britain will help put you in touch with the society of the breed in which you are interested. The address of the Poultry Club is given in Appendix A at the back of the book.

When the commercial hybrids, bred for intensive conditions, first appeared, it was thought that they would be of little use to the less intensive, domestic-poultry unit. Like many a first impression, it was wrong, and backyard hybrid flocks have flourished in all parts of the country.

Unfortunately for our purposes, the breeders do not issue separate production figures for their birds in domestic units. The ones they do release on production, weight, feed consumption and livability are all based on performances in intensive units, but there is no reason why these figures should not be matched and beaten by the domestic producers.

Research and experience in the field has shown that, on average, bird performance declines marginally as flock size increases. Individual bird output will be less from a 100 000-bird unit than a 20 000-bird flock. Similarly the 5000-bird flock will produce more eggs a bird than the 20 000 flock. It follows, therefore, that the six-, twelve-, twenty-five- and fifty-bird flocks should have a field day when it comes to output and many backyard producers who keep records can claim to average 300 eggs a bird.

But to give you an idea of the potential, the performance data in Table 1 is taken from the brochures of most of the major hybrids in this country. The companies are saying, in effect, this is what our birds can achieve under intensive conditions. But before you embark on a study of the performance of the various strains a word of explanation on the terms used at the heads of the columns. *Hen housed average* is obtained by taking the number of birds placed in the house at the start of the laying year and dividing it into the total number of eggs produced during a fifty-two week period.

Percentage refers to the proportion of Large and Standard grade eggs produced by that strain during a laying year.

Feed consumption refers either to the total feed consumed during the laying year or to the amount of feed that the bird needs to lay one dozen eggs.

Weight refers to the bird at the end of lay before it is killed and dressed (i.e. plucked and made ready for the oven).

Livability is another percentage figure referring

Table 1

	Hen housed average	Percentage Size 1 and 2	Percentage Size 3 and 4	Weeks 30–72	Dozen eggs (kilograms)	Weight at 72 weeks (kg)	Percentage livability during lay	Plumage
Brown								
Babcock B380	260–275	31	54	116–121[1]	1.8	2.22–2.36	94	Light brown
Dekalb Amber Link	260	34	53	45–50	2–2.3	2.3–2.4	94	White with brown flecks
Hisex Brown	275	43	49	44.4	1.94	2.2	95	White with brown flecks
Hubbard Golden Comet	255–275[2]	34	31	42–43.5	2.55	2	92–94	Light brown, white under feathers
Ross Brown	270		51 over 65g	42–45	1.9	2.05	94	Brown
Ross Tint	270		51 over 65g	42–46	1.9	2.2	94	White flecked brown
Warren	260–280	30	57	45.36	2.04	2.3	95	Red with white under feathers
White								
Babcock B300V	290–295[2]	10	57	110–113[1]	1.8	1.72	94	White
H & N Nick Chick	260–290	20	70	113	1.8	1.79	90–95	White
Hisex White	279	30	55	41.75	1.8	1.79	90 plus	White
Ross White	275		47 over 65g	41–45	1.9	1.8–2	90	White
Shaver S288	265–285	17	53	112	1.9	1.83	90	White
Shaver S585	260–280	30	50	116–128[1]	1.85	2.38	93	White

[1] These figures refer to the bird's daily intake of food expressed in grams.

[2] In the case of the Hubbard Golden Comet, this is egg production to 74 weeks of age and 76 weeks of age with the Babcock B300V. All others are to 72 weeks of age.

to the number of birds that should on average survive the laying year. A figure of 95 per cent means that five per cent of the birds can be expected to die during the fifty-two weeks under intensive conditions.

Plumage is merely the colour of the feathers.

When buying-in stock you want to be sure that it is the right age claimed by the breeder or rearer. Nine times out of ten it will be, but it is that tenth time you have to be wary about.

Although point-of-lay is a handy term, it is not precise enough to use when buying. Better still to go for birds at a certain age—day-old, twelve weeks, sixteen weeks.

Day-old is used to cover the first 48 hours of a chick's life. At this age it will have a good covering of fluff, smooth on the top of the head, and no indication of a tail. The eyes will be coal black and shine like beads. Height of the chick, 4 in.; weight about $1\frac{1}{2}$ oz.

At twelve weeks the birds will have a full set of feathers and the still bright eyes will be fully coloured. Comb and wattles will be small, but yellow rather than red. There will be no difficulty in telling the girls from the boys at this stage. The males will have a bright red comb and large red wattle. Breasts of the females will be plump. If they are not, send them back—thin birds rarely catch up to make good layers. Weight will vary according to strain from 2 lb to $3\frac{1}{2}$ lb.

At sixteen weeks combs and wattles will have started to develop and to get some colour. Eyes will still be bright, breasts will still be plump. Close-fitting feathers will have a gloss. If the rearer has stocked his birds too closely or not fed them adequately, there will be bare patches on backs and not much tail. Weights will be between $2\frac{3}{4}$ lb and 4 lb.

One of the best guides to age is the make-up of the breast bone. It changes gradually from gristle to bone over an eight month period. At day-old only the front

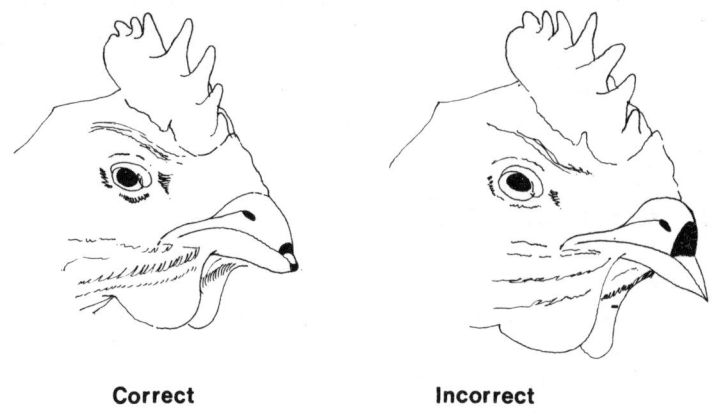

| Correct | Incorrect |

Fig. 10 The correct and incorrect way to debeak laying birds.

is bone and this will spread as the bird ages. At ten weeks it is half bone, half gristle. At sixteen weeks a quarter is gristle. At eighteen weeks the gristle is reduced to an inch in length.

If you are buying-in pullets at eighteen weeks or so do not be alarmed if they appear to have lost the tip of their beaks. Many strains have their beak trimmed before they leave the hatchery to prevent problems with feather pecking and cannibalism at a later stage. It is a routine operation on the day-old bird, but it has to be precise. Remove too little beak and it will grow again. Take too much and the bird will not be able to feed properly. (See Fig. 10.)

So check your new arrivals and see that the job has been done properly. It should certainly not be more than a third of the beak, measuring from the tip to the nostrils. Any more than that and you should reject the bird.

While checking the beaks, place a finger on the nostrils and press down. If a spot of water comes out, the bird has a cold and again is a case for rejection. In fact, check for any sign of disease.

4 Feeding

What do I feed?

Home produced eggs would be cheap indeed if we could get away with feeding the chickens on house-hold scraps, morsels and precious little else. Although they may survive this diet, eggs would be few and far between.

The problem is that modern chickens, although prolific layers, need specially prepared feeds that are packed with protein, energy, vitamins and minerals. They have small appetites so all the 'goodies' have to be packed into a small amount of food. Scraps are bulky as well as unbalanced and will just fill the birds up with the wrong things. But, as we will discuss later, kitchen scraps wisely used can form a very useful supplement to bought-in rations helping to reduce the feeding costs.

To understand her needs it helps to think of the hen as any other highly productive farm animal. She is a machine that converts unpalatable vegetable proteins and energy into tasty products like eggs and meat. But she has to do this over and above maintaining her own body in good condition and she cannot make something out of nothing. You only get out what you put in.

To put it in perspective, if a chicken lays around 250 eggs in a laying year (it may well be more) that amounts to about 30–35 lb of egg material—mainly protein material—bound up in a lot of shell. Even allowing for the fact that the egg is 60 per cent water, she still turns out 10–12 lb of pure dry protein every year in the form of yolk and white, so it follows that this protein must first be put in.

One of the beauties of today's chickens is that they are pretty efficient at converting the food they eat into eggs. There are all sorts of ways that birds can waste food, apart from flicking it all over the floor. They can convert it into unnecessary body fat, as many of we humans do, burn it up in exercise or use it to keep warm and replace lost feathers. But the major international breeders who have developed these birds have been careful to make available only those which are able to avoid these wasteful processes to a large extent. The birds concentrate most of their energies into maintaining their bodies in good trim condition and producing eggs.

To give some idea of the efficiency with which they work most of today's layers can turn out a dozen eggs on about $4\frac{1}{2}$ lb of feed, as well as maintaining themselves. Mind you, it has to be pretty good quality feed. Another advantage that today's poultry keeper has over the 'pioneers' is decades of research into the chicken's feed needs. Years of high-powered scientific research can now be boiled down to the recommendation that the chicken needs a balanced diet of protein, energy, vitamins and minerals. The protein provides the 'building bricks' for her body and the eggs she produces. The energy enables her to carry out all the essential functions of life.

The feed compounders, including giant international companies, such as BOCM Silcock, Spillers and RHM, down to the small country compounders, have simplified the feeding job by manufacturing ready-made, complete meals-in-the-bag that can be bought from any corn merchant.

Having located a suitable supplier in the *Yellow*

37

Pages or local paper, you just pop in and ask for some 'layers' mash'. Purchase it in $\frac{1}{2}$ cwt sealed paper bags for convenience. Failing this, most pet stores will supply layers' mash loose but it is far more expensive to buy this way and more difficult to store. Even when bags have been opened store them in a dry, vermin-proof bin. A plastic dustbin with a tight-fitting lid will make an ideal food store.

Mash, crumbs or pellets?

When buying these layers' rations you may, in fact, be offered a choice of three textures—mash, crumbs or pellets. They are all basically the same except the last two are compressed versions of the mash. Stick with the mash—it is cheaper and less likely to lead to vice problems (see Chapter 6).

How much do I feed?

If you decide to stick to the ready-made layers' mash for convenience and not bother with kitchen scraps, feed it dry, in waste-proof troughs or preferably in self-filling tube and pan feeders (see Fig. 6, page 21). These ensure that feed is available at all times. The thing to watch is that they are eaten out fairly regularly so that there is no build-up of stale feed at the bottom. And, of course, keep the feed dry.

A useful guide on how much mash your birds will consume is that one layer eats between 4 oz and 5 oz a day. These figures take in most situations but if you can control feeding at the lower end of the scale and still get plenty of eggs then do so. In a laying year this will amount to about 1 cwt for each bird. Six birds on the other hand will eat this 1 cwt in two months, so programme your feed purchasing to ensure that there are no gaps when you are without feed. Better to have some in reserve for it keeps well in sealed bags in a dry place.

What about a scratch feed?

A scratch feed means precisely what it says. Food, usually corn, is scattered on the floor so that the birds scratch about to find it. In a way it is an unnecessary part of the bird's diet, but it keeps the birds occupied and prevents boredom. This in turn discourages vices such as feather pecking (see Chapter 6) and it gives them much needed exercise.

Avoid giving too much grain as it is rich in starch which will not only make the birds fat, but also upset the balance of some of the essential ingredients in the mash discussed above. So certainly give no more than 1 oz of grain for each bird a day. Half this amount will still serve a useful purpose—grain is an expensive material to chuck around. For convenience, weigh out the required amount into a suitable container, mark the level on the side and use this as a guide for subsequent feeds.

Grain, like the mash, can be obtained from a corn merchant. Go for a mixture of wheat and maize as oats are expensive and not as valuable. Scatter it by hand in the run if it's not too muddy or wet, or on the floor of the house. If you have only a small run stick to the house floor.

Can I feed household scraps?

There is no doubt at all that household scraps are a valuable way of cheapening the cost of producing your eggs. But they do complicate the feeding job and require some preparation.

Bearing in mind that all the time you are aiming to give a balanced meal, there are a few points that must be understood regarding scrap feeding:

1. *Do not* imagine that everything that you cannot eat is 'all right for the hens'. On the same lines, do not use feed that is unfit for human consumption.
2. Birds *cannot* tolerate much salt, so plate scrapings

that are contaminated with a lot of salt should be avoided.

3. Uncooked potato peelings are useless. So, too, are orange, lemon and grapefruit skins, coffee grounds, tea leaves, banana skins or decomposed meat or fish.

4. Green vegetables or roots should not be included in any quantity as they are far too bulky. By all means feed some raw greens, but separately.

Having said this, the list of eminently suitable scraps is still long. It includes such things as bread crusts, buns, cakes, pies and pastries that have become stale or which are unpalatable through, say, faulty cooking (chickens will polish off slices of burnt toast); potato peelings (cooked), apple peelings, cheese rind and scraps, and leavings of virtually any other kind of food. Incidentally you *must* pressure cook or boil the mix in the minimum quantity of water before use. But don't make it too wet—just 'mushy'.

Neighbours, local schools, hospitals, etc. are all good sources of waste scraps if you have not enough from your own kitchen. But you are obliged to get a licence from the local authority before utilising these scraps.

The complications mentioned earlier arise from the fact that the scraps cannot just be fed alongside the dry mash, for the diet will become unbalanced again. Because of their high water content, most household scraps have only a quarter of the feeding value of the normal mash fed to layers. Four ounces of cooked potato for instance would only substitute for about 1 oz of layers' mash. Hence feeding scraps straight merely serves to fill the bird up and prevent her from taking in the mash which contains all the essential proteins, minerals and vitamins. She is no fool—she will go for the most attractive dish which must be the boiled scraps!

So forget all about the dry mash feeding and think in terms of balancing the scraps with an even more nutritious food. Special preparations available for this are called 'grain balancer rations'. A high protein mash, containing 20–22 per cent protein, is equally good.

Smaller quantities of these rations will be needed and with only a few birds it may be difficult to measure out the quantities the birds require. If this is the case, it is possible to improvise by buying ordinary layers' mash, adding 1 lb of fish meal, or meat and bone meal, to a 1 cwt sack along with 5 oz of a good poultry vitamin and mineral mix. If the vitamins have not been added, give a table-spoon of cod liver oil to each 4 lb of mash.

On no account should these protein rich mashes be fed dry. Apart from anything else it is wasteful. The procedure is to mix some in with the mushy boiled scraps to dry them off. Amounts are arbitrary, but as a rule of thumb guide ensure that the scraps never bulk more than the mash. Once this has been prepared it can be fed in a suitable wet mash trough (see Fig. 5). At the same time some of the protein rich mash should be dampened to a mealy consistency—never sloppy —and fed *ad lib* in a separate trough.

When should I feed?

If you stick to dry mash feeding, this should be made available continuously, save for the occasions when you let the birds eat right out. But if you start continuous feeding do not suddenly leave the birds without feed for long periods.

Grain feeds are best given in the afternoon.

With scraps and wet mash, the morning is probably the best time for feeding although this is not critical.

It can be given at any time providing it is at the same hour each day. Chopping and changing the times will only upset the birds. They will take much longer to finish up the wet mash, sometimes up to three to four hours, so don't rush them. Just ensure that they receive sufficient, based on 4–5 oz each a day and if they finish it up too quickly then give a bit more. You will soon learn from the appearance of the birds and the egg production if you are underfeeding. But don't leave uneaten wet mash around in the troughs overnight.

The amount required—about $1\frac{1}{2}$–2 lb a day for six birds—can be weighed out in a saucepan or bucket and this used as a guide for the next feeds.

You may like to check in the first few weeks whether all birds are getting their fair share of feed. It is often the case that weaker birds or those lower down the social order are prevented from eating by their stronger sisters. This shouldn't be a problem if sufficient feeding space is given, but to be on the safe side check the birds when they have gone to roost in the evening. Just feel the crop with your fingers. It should feel nicely plump with food—about as round as a dessert spoon. If it is almost empty with only a few scraps, then one may safely assume that the bird is being bullied or is ill. A little individual attention to get her over the first few weeks will not go amiss.

Can I feed green vegetables?

Like kitchen scraps, green vegetables are a very useful part of the diet, only this time they should be offered separately from the mash. Cabbage leaves can be tied up or placed in a string bag for the birds to peck at. This will also provide some vitamins and add a deeper colour to the yolks as well as providing a distraction for the birds. Do not throw cabbage leaves or greens onto the run since they soon get soiled and unpalatable. Once those you have hung up become old and wilted they should be replaced with a fresh supply.

Do I need to feed grit?

In the absence of teeth, chickens make use of small amounts of grit which are held in the gizzard and help to grind the food. Grit needs to be supplied in the form of flint or grey granite. Only small quantities are required, say, $\frac{1}{4}$ oz for each bird each month, since it is retained in the gizzard for long periods. Note that grit comes in various sizes—for layers you need 'hen-sized'.

The hen also requires large amounts of calcium for shell formation and although layers' mash or the grain balancers are well fortified with calcium it pays to feed an extra source in the form of a soluble grit, such as limestone granules or finely ground oyster or cockle shells. Whatever you use can be fed in the same vessel as the flint grit and the mixture offered at the rate of about $\frac{1}{2}$ oz a bird a week.

Don't forget the water

A laying hen needs plenty of water—up to at least half a pint a day—and without a regular supply her egg production will certainly suffer. So the supply must be plentiful and clean. In frosty weather it should not be allowed to freeze—warm water given throughout the day will help to prevent this. In summer, drinkers should be replenished frequently to ensure that the water is fresh and cool.

Beyond this, all that needs to be remembered is that all drinkers and feeders should be kept scrupulously clean—a regular weekly scrub in hot soda water, followed by disinfection will help keep disease at bay.

Finally, don't waste feed

It is so easy, particularly with a small flock of birds, to forget that feed is so expensive and that one of the objects of keeping chickens is to provide cheap eggs. So avoid wasting feed—don't let it go mouldy either in the trough or in the storage bins. Avoid overfilling the feeders and if a lot of food is being flicked on the ground remove the feeders for a few hours until it is eaten up.

5 Management of Your Flock

If you are buying pullets at point-of-lay (about eighteen to twenty weeks) the best time to take delivery is early in September. These birds will have been hatched late April/early May and will be coming into lay when the price of eggs in the shops is starting to rise.

It follows that hundreds of other poultry keepers will be thinking on the same lines as yourself and the demand for pullets will be high. So make sure you are well placed in the queue and place your order with the pullet supplier early. If you are able to light your flock during the months leading up to Christmas, when natural daylight is short (more about this on page 50), and provided the pullets are of a good laying strain and well reared, they will lay without a break for twelve months at least. Buy an earlier hatched flock and there is a possibility that they will lay for three or four months and then go into an autumn moult just when you need the eggs. This used to be a very real danger with the pure lines but it is not so common now with the modern hybrid strains.

Assuming the chicks were hatched in May, the first eggs will start to appear about the middle of October. If the end of November comes and some birds have still not performed, the flock would be better off without those particular members. Birds which take a long time getting into the habit, never turn out to be good layers. So be ruthless, cut your losses and kill and prepare the bird for the oven. Take consolation from the thought that it will give you a Sunday lunch at least, even if it did not provide a supply of eggs for the breakfast table.

Who's for the pot?
Culling of birds failing to come up to their potential must be a constant feature of poultry keeping. There is no room for passengers either in the back garden or in the large scale commercial units. They eat expensive food and if they don't pay their way by producing the goods, out they go. It is a hard world.

Later on in the laying year, the same rule applies. If the birds stop laying before reaching fifteen to sixteen months of lay, out they go and you again get chicken for lunch. (See Chapter 8.)

During the laying year there are a number of visible signs, apart from the eggs in the nest, to tell you whether the birds are continuing to come up with the goods. The most obvious feature is the comb. When the bird is laying at full capacity, the comb will be red, large and warm to the touch. The first sign that she is going out of lay is a bluish tint on the tip of the comb and it will feel cool to the touch. If the supply of eggs has dried up altogether, the comb will be cold and have a powdery look.

A bird in full lay will also have a full, large, soft abdomen, a large, open moist vent and a wide gap between the pointed, pelvic bones on either side of the vent. The gap should be large enough to take at least two fingers. There should be $2\frac{1}{2}$ in. between the pelvic bones and the back end of the breast bone (see Fig. 11).

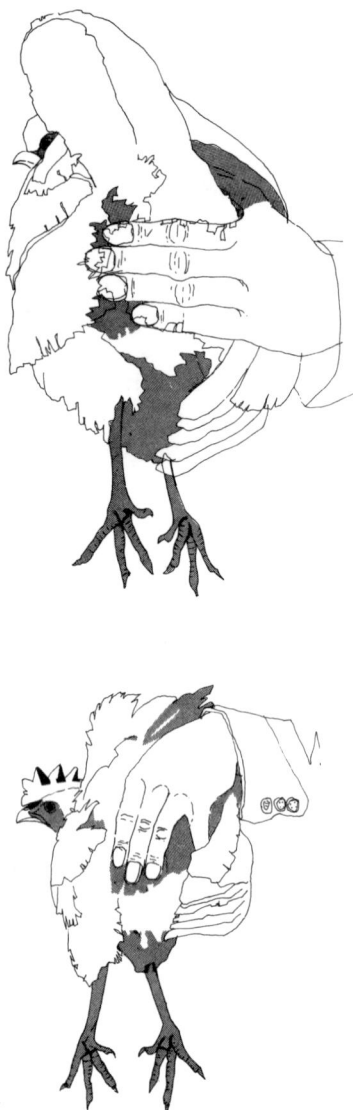

Fig. 11 How to tell if a bird is in lay.
This hen is in full lay. Four fingers can be placed between the
pelvic bones and the end of the breast bone. If there is only a
two-finger 'capacity' the bird is not laying.
A second indication that a bird is in full lay is a three finger span
between the pubic bones. This gap closes up when the bird is
out of lay, leaving room for just one finger.

Out of lay, the abdomen will be small, hard and leathery. The vent will be small, dry and puckered. The pelvic bones will be close together with room for one finger only. The distance between the pelvic bones and the end of the breast bone will be down to 2 in. or less, so take a ruler when checking the laying status of your flock.

Skin colour is also a guide to performance in yellow skinned birds. The colour fades as the year progresses and after twelve months of lay will be white. If a member of your flock still has a yellow skin, she has not been pulling her weight.

The earlier a bird starts laying the better, for you can rest assured that those that produce the first eggs will still be performing in twelve months' time when some of the late starters have stopped.

Get to know who are your best birds so that they can be taken on for a second year. Keep tabs on them, literally, by ringing the first ones in lay and checking their performance. If you are not able to ring the birds, inspect them closely the following September and keep those with full, red combs, moist vents, faded skin and all the other signs of a champion layer.

Finally there is trap-nesting to give the closest check on how well a bird lays. These are nests into which the bird enters easily enough but once she has entered and laid the egg she finds the door will not open the other way. She is trapped until someone lets her out.

Trap-nesting enables an individual record to be kept on each bird, but it also means that someone has to go round every couple of hours to free any trapped birds. It is a job the children in the family might like to take on, but it will probably be at a price.

Hens are at their best if they are left relatively undisturbed and do not have too many visitors. Make a point of getting to know them when they first arrive. This applies to any other member of the household who is going to have any dealings with them.

Start by offering titbits of food and get them to eat out of your hand. It may take a little time, particularly with the lighter hybrids who tend to be shy, but it is all part of the business of establishing a good relationship. Gentle handling and stroking will also help.

When you are in the house or run, avoid any sharp, sudden movements. If you bang your head on the run resist the urge to let out a bellow. Suffer in silence and your birds will lay all the better. Keep well-meaning friends out of the house. They can inspect the birds from outside and keep their yapping dogs and toddlers at a distance.

While we are not suggesting you change your clothes when collecting the eggs, or doing any of the other management chores, it will help if you avoid wearing gaudy, highly coloured materials. A floral dress flapping about in the hen run can be guaranteed to start a panic. A flapping feed sack can have the same effect.

A tap on the door before you open it will warn the birds of your arrival and prevent any panic when you enter the house.

Birds that go broody

Broodiness was at one time the scourge of the poultry industry—backyard and commercial. Hens would go broody at the drop of a hat, but the trait has been more or less bred out with the development of the modern hybrid. The widespread use of cages in the commercial industry has also helped.

Left to their own devices, hens will lay a clutch of eggs and then sit to hatch them. Even the domesticated bird, unless she is caged, will seek out the odd corner for a nest and think in terms of raising a family. Old habits die hard, but if management is good, broodiness can be kept at a distance.

Good management in this context means correct feeding, well ventilated houses and nest boxes, and frequent egg collection. Make sure the rations are correctly balanced with the right proportion of grain and do not continue to feed growers' mash after the birds have come into lay.

Stuffy, over-warm houses and nest boxes will also bring on the broodies, so leave ventilators open in the summer. Although nest boxes should be draught free, a balance must be struck between a howling gale and a broody-inducing fug.

One of the surest methods of turning a hen broody is to leave a few eggs on the nest. In fact china eggs are specially made for this purpose.

The first sign of broodiness in a bird is that she will stay on the nest when you approach. Put a hand near her and she will fluff her feathers and peck at you. The old timers approached a possible broody with palm facing, on the basis that a pecked palm was less painful than a piece out of the back of the hand.

Hens which are about to lay an egg will behave this way, so the only sure sign of a broody is that she stays on the nest all night and the nest stays clean. A bird that is just too lazy to roost at night is also too lazy to perform her natural function on the litter floor or on a perch and the nest will be fouled. But the broody, like any houseproud mum-to-be, will keep the nest spotless.

Having found your broody, how do you break her of the habit? If you can catch her early enough she will be back in lay within nine days at best, twenty days at worst. To achieve this you need a broody coop or cage in which to keep her for a spell (see Fig. 12).

Once in the coop, give her plenty of food and water

49

for three days. If she still goes back on the nest after that it's back to the coop for another spell of solitary.

The alternative is 24 hours with no food but as much water as she wants. After a night and a day on this treatment, let her out in time for food and drink before lights out. Again it is a question of repeating the dose if the bird still makes for the nest.

Let there be light

Lights in a laying house will not increase the number of eggs your hens will lay, but they will smooth out the pattern of production and give an even supply throughout the year. In nature, spring is the natural time for laying eggs. The days are getting longer and the increasing amount of daylight stimulates the bird to mature sexually and produce plenty of eggs. This is the reason for the spring flush of eggs and a consequent drop in prices early in the year.

But once the middle of June is passed and the day-length shortens, egg production declines and there is a relative shortage of eggs in the autumn resulting in high prices. Obviously, no backyard egg producer wants to be caught with plenty of eggs in the spring, and then be faced with having to buy in eggs from the shop in the relatively expensive days of autumn.

The answer is some form of lighting in the laying house that will extend the laying day when natural daylight starts to fade. As soon as the longest day has passed, lights should be introduced to maintain the peak day-length of seventeen or fourteen hours. In July it will be a few minutes only, lengthening through August, and by mid-December the lights will be on for ten or seven hours when natural day-length is reduced to seven and eight hours (see the lighting schedule in Appendix E).

Even at present prices, the load on the electricity

Fig. 12 A home-built broody cage in which the offending bird should be kept for three days with plenty of food and water. An alternative broody-breaking system is short and sharp—24 hours in the coop with no food, but ample water.

bill need not be astronomic and will soon be repaid by the benefits of autumn eggs. A 15 watt bulb will be quite enough for a 6 ft × 4 ft house. If electricity is not available or practical, a paraffin lamp or a lamp run off bottled gas will be adequate.

Fitting a hen house with electric light is a skilled job so, unless you have special skills in this direction, call in an electrician to carry out the work.

Once you start to light your layers you will find they have to be switched on and off at the most inconvenient times—usually in the small hours of the morning. The answer in the commercial world is a time clock, which can be expensive for a domestic unit. It is surprising, however, what can be achieved with a little ingenuity. Alarm clocks (the bell suitably muffled or removed) with a piece of string attached to the light switch and the winding key are not unknown and are surprisingly effective.

Ingenuity is also called for to overcome the absence of twilight with artificial light. Once the lights do go out, the birds will be stuck on the floor and stranded since they do not share with cats the ability to see in the dark. The answer is a false twilight to warn them of advancing night and give them about fifteen minutes to make for the roosts. This can be achieved by shading the existing bulb or by switching to a dimmer light. Another answer is to give them the extra light at the beginning of the day so that they get the natural dusk in which to roost.

When the feathers start to fall
In the natural sequence of events a bird will start to shed its feathers in order to get a new coat before the winter. Normally, egg production stops as the moult starts, but it can vary from bird to bird. Some will not come back into lay until the process is completed, others will lay a few more eggs and come back into production before it is completed.

March and April hatched birds, which come into production in September, will lay for twelve months before moulting. October to February hatches may go into a partial neck moult during the laying year. The first feathers to fall usually come from the head, neck, breast and body. The final ones to go are those in the wings and tail.

A good stockman will be quick to see the first feathers of a moult. If it is not the intention to keep the birds for a second year, they will be killed and on their way to the lunch table in a trice. Again it is a question of not wasting feed where no more eggs can be expected.

The time out of production will vary from bird to

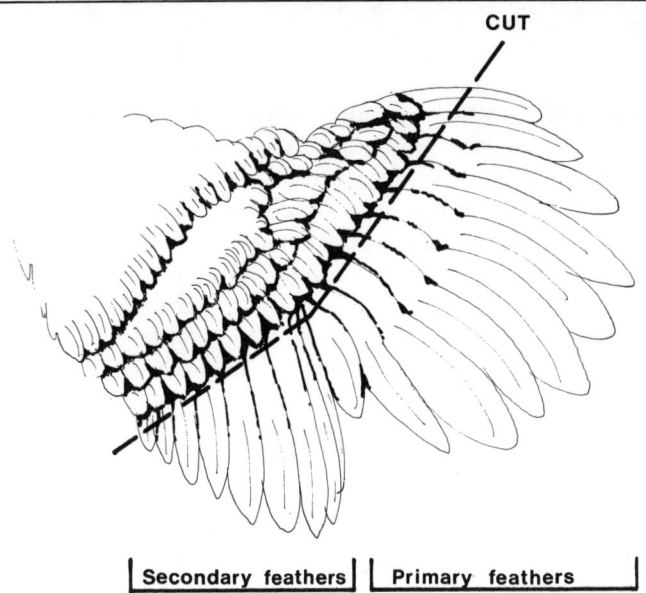

CUT

| Secondary feathers | Primary feathers |

Fig. 13 Chickens are not great aviators, but it may be necessary to clip one of their wings to prevent them flying out of the run. Get a friend to hold the bird with a firm grip, and carefully extend one wing. As he holds the wing open, cut the primary and secondary feathers as indicated—as near the base as possible. Remember to do one wing only. That will be enough to throw any attempts at flight off-balance.

bird. Some will go off lay for six weeks or so, others will take double this time. It is part hereditary and part environment. Those in a well protected environment will soon be back in action. Those on exposed sites will take longer. As a general rule, however, the poor layers moult early, the better ones leave it late.

Moulting is all part of that custom peculiar to females of many species of wanting a new coat for the winter. Autumn is the natural time for shedding feathers, but any extreme experience which upsets the birds is likely to start a moult and stop them laying. Poor feeding, a fright, or an outbreak of disease will trigger-off a moult out of season.

It is possible to use this feature to advantage and

deliberately put birds into a moult. As we have seen, March to April hatched birds will be going into a moult just as egg prices are beginning to rise. If the birds are left to moult in their own time, you will be collecting eggs during July and August when they are relatively cheap in the shops. But as soon as the weather gets a bit colder and the national level of production drops, your birds will start to moult and the supply of eggs will dry up just when you need them most.

By causing the feathers to drop early (July and August) the flock can be back in lay by September or October and producing a better quality egg than if it was allowed to moult naturally. Second-year birds produce far more extra-large eggs than in the first year, but shell quality tends to deteriorate. The mechanics of producing the shells are not always capable of coping with the extra size and the result is a relatively high proportion of thin shells leading to cracks and breakages.

Supporters of force moulting maintain that the technique improves calcium output and the interior quality of the egg, leading to stronger shells and thicker whites which will stand up well in the frying pan.

The shock treatment which starts off the force moult is similar to that given to break broodies, except that it applies to the whole flock and not just one bird. The most common method is to keep the birds and house shut up for an entire 24-hour period with no feed and no water. Give them a little water after 24 hours and a hopper of bran or oats after 48 hours. These are the maximum limits allowed by the Ministry of Agriculture's Code of Welfare for layers and they may not always be sufficient for brown-egg layers. It is easier to get a white egger into a moult, but some of the heavier strains of brown bird are more stubborn and require more drastic treatment. This means longer without feed and water, but both courses would be outside the welfare code.

Force moulting can be a difficult operation for a beginner and it takes a degree of stockmanship to get the birds back into lay within a reasonable time. Feeding has to be carefully controlled and, if you are only doing part of the flock, separate accommodation will be necessary. But if you are prepared to take the trouble and have the facilities, there is no reason why it cannot be done as successfully in a back garden as in a commercial unit. Indeed it should be easier in many respects in a domestic unit where fewer birds are involved. But do not be disappointed if it does not go exactly according to plan the first time. It boils down to management and stockmanship and these attributes are what divide the successful from the unsuccessful poultry keepers. They can be acquired only with time.

The autumn moult is not the only time birds shed their feathers. They spend much of their early life moulting and have had four changes of feather by the time they are twenty-eight weeks of age. Chicks are hatched with a covering of fluff which lasts until the first feathers appear—the wing feathers first. It does not take long to complete the set, but the bird is in a state of almost continuous moult until a few weeks after point-of-lay. Do not worry, therefore, if the run and house are littered with feathers—they are being replaced.

Ministry watch on welfare
The welfare code, referred to in the section on moulting, was introduced by the Ministry of Agriculture in 1969. It is intended as a guide line for both commercial and domestic poultry keepers and as such carries no legal force. It is not a question of being

liable to court action if the code is broken in any way. But any transgressions of the code could be used by the prosecution in a court case.

Much of the code relates to the commercial industry, for which it was primarily designed, but there are a number of points which the domestic poultry keeper would do well to note. In the introduction, the code speaks of the importance of recognising the signs of impending trouble and taking suitable avoiding action. The signs of good health are listed as alertness, clear bright eyes, good posture, vigorous movements, active feeding and drinking, clean and healthy skin, shanks and feet. Signs of ill-health in layers are a drop in egg production and changes in egg quality, such as shell defects.

In planning new houses, consideration should be given to building materials with a high fire resistance and electrical installations should be planned and fitted to minimise fire risk, says the code. Nest boxes and roosting areas should not be so high above floor level that birds descending from them are injured. Cages should be designed and maintained so as to avoid injury or distress, and be high enough to allow the birds to stand up normally.

Birds should be protected from draughts and not be exposed to strong, direct sunlight long enough to cause heat stress as indicated by prolonged panting. There are also warnings about panting due to overheating of young chicks and about prolonged huddling which indicates that the birds are not warm enough.

On lighting, there should be provision for a period of darkness in each 24-hour cycle, indicating that lights should not be kept on throughout the night. There should also be enough light to enable all birds to be seen clearly when they are being inspected. All equipment should be inspected regularly to see that it is working—this includes feeding and watering equipment.

The code also recommends a series of minimum space requirements. For layers on deep litter it recommends that the floor area, including any slatted or metal mesh area and area occupied by equipment, should be enough to allow not less than 1 sq. ft for every 3 lb liveweight. This means that a 5 lb bird should have 1.66 sq. ft minimum. The recommended allowance for birds over 7 lb liveweight is 1 sq. ft for every 3.5 lb.

On the rearing of birds for laying, the recommendation is 1 sq. ft for 4 lb liveweight. It is related to the size of the birds which will be in the house at the end of the rearing period.

Recommendation for layers in cages is that each bird should have a minimum trough space of 4 in. and that where there are three or more lightweight birds in a cage, the area should be at least 1 sq. ft to each 8 lb liveweight—heavier birds get 1 sq. ft to 9 lb liveweight. In two-bird cages, it is 1 sq. ft to 6 lb liveweight and in single-bird cages 1 sq. ft to 4 lb liveweight.

There is also a recommendation for birds in range arks of 1 sq. ft to 8 lb liveweight where the floor is slatted, and 1 sq. ft to 4 lb liveweight where the floor is solid.

To emphasise the point, the code prints in heavy black type that these allowances are given as a guide. If birds are stocked more tightly 'sound management may become the critical factor'. None of these allowance recommendations should prove a handicap either to the domestic or commercial sectors, but it is as well to know what the Ministry regards as minimum.

In a number of additional recommendations for range birds, the code makes the point that stock should not be kept on land which is 'fowl sick', i.e. land which has become contaminated with disease-causing or disease-carrying organisms. The birds should be protected against foxes, other predators, and dogs and cats. There also should be shelter from rain and sun, and windbreaks on exposed land.

In range houses, precautions should be taken to avoid crowding and suffocation when birds are first moved in. They should not be confined for long periods during daylight hours or subjected to direct sunlight during confinement.

Food and water should not be left in a contaminated condition and care should be taken to see that unfrozen water is available in freezing conditions. In the general recommendations on food and water the code says that during the force-moult no bird should be without food for more than 48 hours and without water for more than 24 hours.

The welfare codes cover various types of farm livestock. Code No. 3 is the one dealing with domestic fowls and it can be obtained free from your local office of the Ministry of Agriculture. So get on to them straight away—it is a very useful booklet to have on hand.

6 Protecting the Investment

Predators, rodents and insects

Predators

Before you even start to worry about the host of poultry diseases that *could* attack the insides of your chickens, guard against the outside predators such as foxes and to a lesser extent rats. Without doubt, even if you have never seen a fox in your particular garden, they are near at hand and will be attracted by your chickens.

So first and foremost, if your birds are to be given any type of open run it will need a sturdy wire-netting fence, at least 6 ft high and dug into the ground at least 1 ft around the circumference. This at least will act as a deterrent.

Since foxes generally appear at night it is also essential that the birds are shut away at this time. To leave them roaming free is asking for trouble, so a routine daily chore will be to see that every bird is housed at dusk and the door firmly bolted. This is no great problem for the birds will naturally go inside as night falls but, left to their own devices, they would be out again well before dawn giving ample time for foxes to attack before you are up and about.

In fact, shutting them away at night has a dual function. It also ensures that in adverse weather, or when the run is particularly wet and muddy, they are kept in the clean and dry. It will certainly do them no harm even if they have to stay inside for several weeks. Just remember to provide adequate food and water in the house.

Rodents

Although rats and mice present no particular hazard to the birds themselves, infestations can build up quickly as these animals are attracted by the food, warmth and dark corners around the house. Not only can this present you with a costly loss in terms of the feed they may eat (one mouse will eat only about $\frac{1}{6}$ oz of mash a day, but they rarely come one at a time) but it is totally unacceptable to both your family and your neighbours. There are not many quicker ways of losing your friends and possibly permission to keep chickens than giving rise to a plague of rats in the area!

Rats, of course, can cause structural damage by gnawing through cables, waterpipes and walls. As soon as you see signs get in touch with the pest control officer at the local council.

Mouse-proof buildings. Mice leave the nest when very young which means they can enter buildings through $\frac{1}{4}$ in. diameter holes. So the first thing to ensure is that entry points are blocked off as far as possible. This means fitting vermin-proof flaps on the bottom of the doors. Ensure that there are no holes leading to hollow walls or ceilings or cosy corners in the nesting materials.

Planned control and prevention. Sooner or later mice will get into your poultry house so set up permanent baiting points in the hope of killing any invaders before they can multiply.

There is little point in using anti-coagulant poisons (like Warfarin) against mice because even if they are not resistant, poultry food is so rich in vitamin K, an antidote to anti-coagulants, that they must take in extra large doses before they succumb. One of the best and safest alternatives is a 4 per cent alphachloralose mixture.

Mix 14 oz of whole wheat, 14 oz of porridge oats, 14 oz of canary seed, that has first been treated in an electric coffee grinder for 2–3 seconds to remove husks (if husk is not removed first the poison sticks less well to the seed), and 1 oz of icing sugar; then add 5 oz of corn oil and mix well; finally add 2 oz of alphachloralose and mix again.

This makes 50 oz of bait for less than £1 which is enough for thirty to forty baiting points.

After several weeks, however, this bait can become less attractive to mice as the oil goes rancid. So to make a mix that will last for several months and can be laid in permanent baiting points the same ingredients can be incorporated into wax blocks as follows:

Mix 18 oz of whole wheat, 18 oz of porridge oats, 16 oz of pinhead oatmeal, 16 oz of dehusked canary seed, 3 oz of icing sugar, 5 oz of corn oil and 4 oz of alphachloralose. This bait should then be mixed into 20 oz of melted paraffin wax (20 oz blocks of paraffin wax are available from some of the larger chemists). When set, the wax should be cut into 1 in. cubes which are then ready for use.

These paraffin bait blocks are particularly useful as they can be thrown into places that would be inaccessible for laying ordinary bait. They may be put into suitable containers in poultry houses to provide permanent baiting points. This keeps them free of dust and they remain palatable. An empty tin, with an entrance hole of about 1 in. square cut in it, may be used as a permanent baiting station. These tins can be nailed to woodwork in suitable corners, or weighted firmly beneath houses or in the droppings pits with bricks etc.

If you have difficulty in obtaining the alphachloralose, you should consult the Pest Officer at the Ministry of Agriculture's nearest Divisional Office.

Insect pests

Insect pests, such as fleas, lice and mites, affect your birds directly and seem to come out of the blue, although they occur less frequently in light, well ventilated houses. The most damaging are red mites and fleas.

Red mites. These are in fact greyish in colour, about the size of a small pinhead and live in cracks and joints in the woodwork of the house. They visit the birds at night to suck blood and it is only then that they appear red. Heavy infestations, that are easy to overlook because the insects live away from the birds, can result in anaemia and lower egg production and, in some instances, can deter the birds from laying in the nests.

Heavy infestations in a house may at first be recognized by reddish-black droppings on the eggs. When they appear, check the undersides of the nest-box lids, among the litter and in crevices. Small clusters of the insects may be exposed that scatter when disturbed. Their droppings show up clearly against dark woodwork. Examine the birds at night time when the mites feed.

The best way to prevent them is to treat all the woodwork of the house, especially cracks and joints, with insect killer. Products containing pyrethrum, malathion, dichlorvos and gamma–BHC (Lindane) are eminently suitable and are available as sprays. Carbaryl is particularly suitable against mites and other parasites, and harmless if used correctly. As a precaution, move the birds out, take away any eggs, and empty feeders and drinkers if inside. Give the house a good clean down first and burn the litter.

Spray all parts of the inside of the houses, especially the perches, surrounding areas and the insides of the nest boxes. Another treatment, one to three weeks later, is necessary to kill off insects that may hatch from surviving eggs. To kill off mites remaining on the birds, dust them individually with carbaryl, or one of the other compounds mentioned, before returning them to the house.

Do the job twice a year, in early spring and late summer, and you need not fear the red mite. Combine with the treatment, preferably in late summer if possible, a thorough coat of creosote for the outside of the house.

Fleas. These creatures also visit the hen for food only; they live and breed in dark places in the house, especially the nest boxes. The grubs of the flea live on dirt in the house so a clean house is unlikely to harbour them.

The dusting powders mentioned above will also keep fleas at bay. When spraying, remove all the old litter and spray the floor as well as the inside of the house. Nest boxes should be treated once a year, but not when the birds are laying or the eggs may be tainted. See that you keep the nests clean and change the litter at least once a month. Wash out the nests and sprinkle with a little dry lime on the floor before putting in new litter.

Lice. Lice are possibly the most familiar type of external parasite of chickens and are easier to locate as they live and breed on the bird itself. They probably do the bird little harm but are best eliminated. The bird herself preens and takes dust baths as natural protection, but a proprietary dusting powder containing sodium fluoride should be used.

To treat, take each bird from its perch at night and put a pinch of the white powder under each wing, in the feathers at the back of the neck and on the back, and two pinches in the fluff under the tail and vent. Repeat the process after ten days. The birds can then be left for three months or so before the next treatment. The ten day interval between the two treatments is very important because although the first dose will kill all the live lice on the bird, it will not affect the lice eggs. In ten days' time all the eggs will have hatched, but the oldest lice will not be old enough to have laid again. If a shorter or longer time is left between the two treatments some eggs may be left to start another generation.

Malathion is another useful product for keeping down all insects and lice. A 4 per cent malathion dust can be obtained from horticultural stores and agricultural chemists. If $\frac{1}{4}$ oz of this is mixed with the floor litter for each 8 sq. ft of floor space and a few pinches added to the nest box litter once every two to three months, you should have no trouble with these pests.

Diseases and health

The list of diseases that chickens can catch is so long as to be frightening. Fortunately, the vast majority will not worry the small flock owner. The real problems arise on the large intensive units where diseases can spread rapidly because all the birds are sharing the same enclosed environment. In comparison, the garden flock should be a particularly healthy lot and problems should be few.

The best advice for a small flock owner is—if a bird, or birds, looks sick then isolate it from the rest, keep it in a warm place with feed and water and give a dose of Epsom salts and, possibly, cod liver oil. If it shows no signs of recovery within a few days or gets worse, then put it out of its misery. Paying out for expensive medicines and veterinary bills cannot be justified. There is no guarantee that treatment will work anyway.

The most common complaint of all is a lack of proper food and water. Stick to the recommendations in Chapter 4. Make sure there is a good meal and a drink available *every* day and you won't go far wrong.

If you do have the odd death and the cause is not obvious, or if a number of birds start to die and the rest are in danger, you may feel you need to get an expert opinion on the trouble. Rather than call in a vet, you can send carcases (providing, of course, that they are fresh!) to your local Ministry of Agriculture Veterinary Investigation Laboratory. Or better still, give them a ring first and deliver the bodies by hand. There is a small fee for the service.

The address of the nearest laboratory can be found in Appendix A.

Should you be forced to send carcases by post, make sure you wrap them well in strong brown paper. Include a covering letter carrying details of the history of the bird, any symptoms you noticed, the type of feeding, egg production and some details of your own unit.

What follows is a check list of some of the more common health problems, with some ways to avoid or, if necessary, treat them. It would be unreasonable and unwise to look closely at all poultry diseases. It would certainly be outside the scope of this book for in many cases diagnosis is for the experts.

(a) *Birds become listless and droopy, lose weight and produce watery droppings or whitish or chocolate diarrhoea. This may be accompanied by ruffling of the feathers and huddling to keep warm. Egg production may also be reduced.*

If birds show the above symptoms, check first that they are being fed properly and that there is plenty of water available. If there are no problems here it is possible that they have picked up some intestinal parasite such as worms or microscopic organisms called coccidia. If there is blood in the droppings as well, coccidiosis can be strongly suspected.

With both these diseases, eggs of the parasites thrive in damp, earthy conditions and are picked up from the litter or run. So one way to avoid the problem is to keep the conditions clean. Change the run occasionally or at least give it a rest. Dig it over, treat with lime and let the sun and weather clean it up.

If worms really are suspected—they may even be visible in the droppings—here is one instance that drug treatment can be considered. Since more than one type of worm may be involved, two separate treatments may be needed. The drug phenothiazine can be fed at the rate of 0.2 g a bird, the medicated mixture being given on three successive days. This will control the commonest worm, the caecum worm.

For removal of the large round worm that invades the small intestine, one or other of the several piperazine compounds on the market under proprietary names can be administered with the food. Follow the makers' instructions closely.

Coccidiosis is even more difficult to diagnose with certainty than worms. The one characteristic symptom of a severely diseased bird is that it rapidly loses weight and stands huddled-up with the feathers ruffled. The droppings will be watery and greyish, often containing green bile stains. If you suspect the disease, and more than one bird looks sick, then treat with a proprietary coccidiostat that can be given in the drinking water. Products are also available that can be given in the feed and some large producers use them continuously for birds on litter. Check with your feed supplier.

(b) *Birds go off lay or produce fewer eggs. Some eggs may be weak or soft shelled. The birds seem to have a heavy cold (gasping, hoarse chirping, rattling of the wind-pipe) accompanied by loss of appetite and increasing thirst. Look also for nervous symptoms such as partial or complete paralysis of the legs or wings and unusual movements of the head. Some birds may die suddenly.*

If the nervous symptoms are definitely present, along with other signs, then it is a fair indication—though by no means certain—that the birds have a particularly nasty disease called *fowl pest (Newcastle disease)*. If you suspect it, particularly if birds are dying, there is only one thing to do—you must notify the local Ministry of Agriculture Office. They will send someone along to do tests on the birds to confirm presence of the disease, or otherwise. This is the only poultry disease for which you are obliged by law to notify the authorities—so serious is its potential.

Fortunately, there are some very effective and reliable vaccines available that prevent the disease occuring nine times out of ten. A series of at least three vaccinations is given during rearing to build up immunity—in much the same way as we are given a series of tetanus jabs to make us resistant to that disease. When buying point of lay pullets, insist that you receive a certificate to show that they have been vaccinated, not only against fowl pest but infectious bronchitis and Marek's disease (see later). They will cost a little more than unvaccinated birds but are well worth the extra.

It is common practice these days for rearers to give a particularly strong dose of fowl pest vaccine to birds at between sixteen to eighteen weeks of age. That should protect the bird for the rest of its life—certainly for the next twelve months or so. But to be on the safe side, it will pay you to give them a 'booster' vaccination after they have been in lay for three to four months.

Unfortunately, although you can easily buy the vaccine (your pullet supplier will tell you what to get and where to buy) the smallest pack available is sufficient for 250 birds. And once opened it will not keep for more than an hour or so. So inevitably most will be wasted. It is still worth using it though, for the vaccine will cost less than the price of one bird.

Having purchased the vaccine, follow the storage instructions carefully. It is a live virus preparation and becomes useless in no time if exposed to warmth or sunlight.

Birds are dosed through the drinking water. Before you start, drinkers must be emptied and cleaned with water only, *not* with detergents or disinfectants as these will nullify the effect of the vaccine. Leave the drinkers empty for two to three hours before vaccinating to ensure that the birds are thirsty. Then mix the vaccine according to the instructions. The 250-dose pack should be dissolved in about one gallon of fresh, clean water in a clean plastic bucket. Clean mains water should be used, but if there are likely to be any impurities, such as free chlorine, mix in half a pint of skimmed milk as well. For six birds about half a pint of the vaccine mix should be added to the drinkers. If this is consumed rapidly, add a bit more. Discard the left-over vaccine mix unless a neighbour wants some immediately.

Should the above mentioned cold-like symptoms be present without the nervous condition, *infectious bronchitis* (IB) may be causing the trouble. This possibility, however, is fairly remote providing that the birds received the standard two IB vaccinations during rearing. It only really becomes a problem where birds are crowded together in large numbers in the same house. A mild form of the condition leads to depressed egg production, wet eyes and swollen sinuses, all of which will clear up in a couple of weeks and cannot really be treated anyway.

(c) *Paralysis of the wings or legs without cold symptoms. Birds become thin and lose weight rapidly as they cannot get to the food. A wing may begin to droop, and the bird may fall back on its hocks. Sometimes, very sudden death without any signs of disease.*

These symptoms are fairly typical of a virus disease that used to be the scourge of the industry—*Marek's disease.* It is far less common today, thanks to a live vaccine, given at day-old at the hatchery, that protects birds for life.

The disease, when it occurs, is more common in the rearing stage than in laying birds. Should it occur then there is absolutely nothing you can do but kill the bird, destroy the carcase and hope the disease does not spread. Alternatively send the carcase to a veterinary investigation centre with details of symptoms. If

Marek's is diagnosed then have a word with your pullet supplier about possible compensation.

An alternative condition that results in paralysis of the legs only is *layers' cramp*. Production continues unaffected, the comb red, the bird in good condition and the eyes bright. For treatment the bird is placed in a littered coop, given a teaspoonful of Epsom salts in the water and fed grain for three days only.

(d) *Birds lose feathers early in lay as if in moult. Egg production suffers.*

This early moult is almost certain to be brought on by some adverse factor like faulty lighting, sudden lack of water or food, or a particular deficiency in the diet such as a shortage of calcium.

All you can do is check the factors mentioned in the hope that the fault is located and put right.

(e) *Birds show no specific symptoms other than being off-colour with no interest in feed. May have diarrhoea.*

This could be the start of many things, but suspect first a digestive upset. Simplify the feeding by cutting out the grain feed—which is difficult to digest—and the greens to see if the condition clears up. If it does not, watch for other symptoms.

(f) *A bird makes frequent and prolonged visits to the nest without producing an egg, or leaves the nest and walks round looking most uncomfortable, perhaps straining as if to pass an egg, usually with the tail down.*

Most likely to be *egg binding* where the bird is trying to pass a very large egg or where there has been a derangement of the oviduct.

In such cases, holding the bird's vent and abdomen over steamy hot water for a few minutes, or plunging the bird's rear end into warm water has good results. The bird is then put into a warm, hay-lined box or basket where she will probably lay. But if the problem recurs frequently you may have to destroy her.

If the egg remains stuck, on no account try to hook

or force it out, even though it may be visibly protruding at the vent. Handling the bird very gently, with the forefinger dipped in olive oil or vaseline, try greasing the whole of the vent, pushing the egg very carefully backwards into the passage in order to do this effectively. Then, holding the hen with the back part downwards, work the egg carefully and slowly along the passage until it drops out through the greased vent.

(g) *A portion of the internal organs protrude through the vent, often when the bird is endeavouring to lay.*

Prolapse, as this condition is called, is fortunately not very common for it is pretty unpleasant for all concerned. The most common cause is prolonged, heavy egg production or straining to produce over-large eggs.

The condition may occur with the egg laid or with it still inside. If inside, it must be got away as described in the preceding section under egg binding. Once this has been done it is worth trying to replace the organs. The hands should be washed thoroughly, using disinfectant, and gentle efforts made to replace the parts. Unfortunately, the treatment is rarely permanently successful and it is usually better to destroy sufferers and prepare them for the table. But give it a try first.

(h) *Abnormal distension of the crop region. The swelling may be either soft and mushy or hard to the touch.*

If the swelling is hard the condition is likely to be *crop binding* resulting from bits of straw, hay and other stringy material clogging the outlet of the crop.

It is easier to guard against crop binding than cure it. Examine the birds regularly and the moment any abnormal distension of the crop is noticed, isolate the bird, keep it for a day without food and then re-examine. If the distension persists the bird is almost certainly crop bound and needs treatment.

This is relatively easy if the diagnosis is made early. Using a spoon, give the bird as much warm water as the crop will take. Then, with the bird's head held downwards, knead the crop gently with the fingers for a few moments until the contents are ejected from the mouth. Give a teaspoonful of olive oil and place in a coop. Next morning, if the crop is not empty, repeat the process. Should the treatment fail and the condition persist, the bird ought to be destroyed. There is a surgical operation that will alleviate it, but that is a job for a vet.

Sour crop is a similar condition only this time the crop is greatly distended and the contents soft and flabby, being composed of liquid instead of solids. By holding the head downwards and gently massaging the crop, the liquid content can be expelled. When this has been done give a pinch of bicarbonate in warm water.

(i) *Swelling on the ball of the foot or between the toes which can spread up the leg causing lameness.*

Aptly called *bumble foot*, this condition is caused by a bit of cinder, stone or metal becoming embedded in the foot. If discovered in time, painting with tincture of iodine will often put things right. But if this fails the swelling should be incised and the contents removed.

With a sharp knife, sterilized by immersion in boiling water, cut a small opening in the swelling and gently squeeze or scoop out the contents. Cleanse the wound thoroughly with warm water in which a few crystals of potassium permanganate have been dissolved. Bandage the foot to prevent entry of dirt and isolate the bird for a week or so.

(j) *Birds pecking at each other's feathers, possibly reaching a point where the skin is broken and a bloody hole has been made in a bird's back or side.*

This vice of feather pecking is often the result of over

crowding in small houses or runs, or boredom. Many things can trigger it off but if allowed to continue it can lead to death through loss of blood or infection in the wound. And it will certainly depress production. The odd moulting bird is a ripe candidate for attack, for the short stubby feathers are a big attraction.

At the first signs of feather pecking, apply anti-peck ointments around the wounds. Make your own preparation by mixing together 4 oz of petroleum jelly, $\frac{1}{4}$ oz of aloes and $\frac{1}{4}$ oz of carmine. Then look for the cause. Check that there is sufficient feeding space (4 in. per bird) and some greens hanging in the run to occupy the flock. Check too that the birds are de-beaked. Offer them a dish of meat and bone meal, or partly cooked meat or flank of bacon hung up in the run as an alternative.

Preventing the problems

'Prevention is better than cure'—it's an old and corny saying but it still holds true for all that. The more you improve conditions for your flock and take the necessary disease precautions, the less you will have to worry about ill health.

One of the keys to success is constant observation. This is all part of stockmanship; something you do not acquire overnight. But a keen eye kept on the birds will catch the first signs of many a potentially dangerous condition. A close examination once a week will reveal the presence of lice, fleas and other insects. Make records too of egg production so that deviations from the expected normal pattern will act as a warning that all is not well. Ensure that all birds are getting their fair share of feed. If the odd one is going short through bullying it will become weak and more susceptible to illness.

You will be half way there if you purchase healthy, well-reared and fully vaccinated, point-of-lay pullets from a reputable breeder. A comprehensive list of suppliers is given in Appendix A and there should be one in your area.

Carry out Newcastle disease vaccinations as described earlier and should you decide to rear your own birds from day-old ask the hatchery to suggest a good fowl pest and infectious bronchitis vaccination programme. It will be costly but cannot be ignored.

Cleaning the premises

The house should be rested for at least ten days between finishing an old flock and starting a new one. This gives time for a thorough and essential clean down. All appliances should be removed, the litter taken away and disposed of—possibly on the garden. The last traces of manure can be removed after the house has been given a thorough soaking with water for a day or so.

The house should be scrubbed with water and washing soda followed by a thorough soaking with an all-purpose disinfectant. Leave for a couple of days and then wash again with water and allow the house to dry out. If this can be repeated more than once a year all the better, particularly if there has been infection or infestation about. A coat of creosote painted on the outside each time the inside is scrubbed out will pay dividends.

As far as maintenance of the run goes it should be moved as often as possible to avoid a possible build-up of disease. But if it is fixed then maintain it in good condition by not using it in wet weather if it is not covered, digging it over occasionally and treating it with lime and insecticide. Better still, cover the surface with 6 in. or more of cinders or shale to give a free draining surface (see Chapter 2).

Make sure there is always a supply of fresh, cool water available. The container itself should be washed every day and sterilized with boiling water once a week. Clean the droppings board weekly, taking the manure well clear of the birds and run.

Avoid using hay and straw in the nest boxes as this can lead to disease—stick to wood shavings or sawdust.

For the litter, a mixture of wood shavings, sawdust and straw is ideal. Keep an eye out for damp patches on the floor. If you spot any, remove the source of the damp, dig out the caked litter and replace with fresh material. If you fail to get the litter to 'work', i.e. in raised floor houses, there may be no alternative but to clean out and replace with fresh material at regular intervals.

7 The Egg Harvest

If everything goes according to plan, if the birds are bought healthy and remain healthy, and the feeding and management tips mentioned so far are followed, then each bird should go on to lay between 240–300 eggs over a complete laying year of about 52–55 weeks. That is, a little less than one a day from each bird (see also Chapter 9).

What this means on a daily basis is that if you have six birds you can expect, on average, four or five eggs a day. Twelve hens will produce around eight or nine eggs a day, and so on.

Collection
The first thing to avoid is an accumulation of eggs in the nest boxes. Although it is nice to pop down the garden on a sunny morning to pick up a few eggs, it is important to keep this up when the first flush of enthusiasm has waned and the morning is dark and

wet. Eggs lying in the nests are likely to get badly soiled, become damaged and encourage broodiness. For safety's sake, take the trouble to look in two or three times a day.

Always give the birds warning before you go into the house—tap on the door before entering so that you don't surprise them. Move about the house slowly and methodically. Rapid movements will disturb the birds and those taking sudden fright from the nest are likely to damage the eggs.

Small numbers of eggs are best collected in a bowl or a fibre egg-pack. When larger numbers are involved avoid using a soft container with handles since cracks will almost certainly result. Use a thirty-egg Keyes tray or a rigid basket with some tissue at the bottom.

Storing eggs

Unless you have a large family to feed, do a lot of baking or supply your neighbours, then six to eight eggs a day may prove to be more than you require. There is no need at all for this to be a problem for, providing they are kept in a cool room where the temperature does not rise above 50°F, good quality, undamaged eggs will keep in good eating condition for up to three to four weeks. They are not quite as attractive on the plate after this time but the only real change that takes place is that the white becomes runnier.

Ensure that eggs are stored pointed end down on racks or clean Keyes trays and don't hold onto cracked eggs for too long as these are more susceptible to bacterial invasion. Don't store them next to onions, fish, cheese or any other strong smelling foods as they can easily pick up smells. It is not a good idea to store any eggs in the fridge as they deteriorate rapidly once they are taken out.

How fresh?

Experience will tell you whether you want to eat eggs on the day they are laid or hold them for a few days. It is certainly a nice idea to pop out and collect your own breakfast egg but some people find them too fresh. The whites may be a little thin—they firm up during the first few days and then go thinner again in storage—and sometimes the yolk seems to lack flavour. But you will discover your own preference and will establish a stock rotation that suits you best.

Long term storage

You will find that while the spring and summer months will provide a 'flush' of eggs, supplies in winter will dwindle, unless you are able to provide your layers with artificial light (Chapter 5). Even then, while there will probably still be enough for meals, eggs for cooking will be scarce.

(a) *Storage in water glass*. By far the simplest and cheapest method of storing eggs for periods of up to a year is to put them in water glass solution (soluble sodium silicate). This seals the pores in the shell, prevents the loss of moisture from the inside and keeps out bacteria.

It is obvious that only fresh eggs, unwashed and of the best quality, should be preserved in this way. The shell should not be damaged or soiled.

The sodium silicate can be bought from any chemists and many hardware stores. It is made up domestically by mixing equal parts (by weight) of the powder with water. The viscous solution is then diluted—1 part to 20 parts of water—before adding the eggs to be preserved.

Earthenware jars are the best vessels to use although enamelled pans are suitable. The water glass solution is poured into the vessel and the surplus intact eggs are placed into the solution each day as they are gathered. Place the eggs broad-end upper-most and ensure that there is always about 2 in. of solution above the top layer of eggs.

Since water will evaporate, some form of cover to the pots is a good idea. Occasionally the solution will need topping-up to ensure that the eggs are covered at all times. When they are required, simply take them out and wash thoroughly under running cold water.

Eggs stored in this way are only really suitable for cooking and baking and must not be sold.

(b) *Deep freezing*. A more convenient method of long term storage is in a deep freeze. It offers a range of possibilities and enables cracked but fresh eggs to be stored.

For instance, four to six eggs can be beaten, half a teaspoon of salt or sugar added to prevent thicken-ing, and the mixture packed into a wax container for freezing. Fewer eggs can obviously be treated in this way for omelettes or for baking.

Another possibility is to take as many eggs as there are compartments in the ice-cube tray, break them out and mix as above, pour into the tray, replace the divider and freeze. Frozen cubes can then be stored in the freezer in a polythene bag, each cube representing a single egg.

For even more convenience for cooking and baking at a later date, the yolks and white can be separated. The yolks are mixed as above with a little sugar or salt and frozen in small cartons or an ice-cube tray. In this case one tablespoonful is considered the equivalent of one yolk.

The whites can be frozen as they are with no beating or additions. Here two tablespoonfuls are equivalent to one white.

Eggs stored in the freezer will last at least six months. Just remember to label the packs, marking whether salt or sugar has been added.

Grading and selling the eggs

Should you find yourself with more eggs on your hands than you can reasonably use or store, you may consider selling the surplus. Under the present laws of the land you are allowed to sell them to neighbours or other households. You are not allowed to supply shops with eggs for resale unless you erect and equip a fairly sophisticated packing station which then has to be inspected and granted a special licence by the Ministry of Agriculture. This would not be worthwhile unless your bird numbers were up in the thousands.

It is best to sell by price at a figure agreed between yourself and the purchaser. It gets complicated if you consider selling by grade for you need an accurate weigher that has to be approved and regularly checked by a Weights and Measures Inspector. Hand weighers are, however, available for around £25 and you may wish to purchase one for your own interest.

In November 1977 the UK changed from the imperial system of grading eggs which gave five named sizes (Large, Standard, Medium, Small and Extra Small) and adopted the EEC system based on seven metric sizes at seven-gram intervals. The seven sizes are:

> Size 1—70g and over
> Size 2—65g to 70g
> Size 3—60g to 65g
> Size 4—55g to 60g
> Size 5—50g to 55g
> Size 6—45g to 50g
> Size 7—under 45g

Egg quality and faults

It is worth being prepared in advance for the sort of problems that can arise since you will almost certainly encounter abnormal eggs.

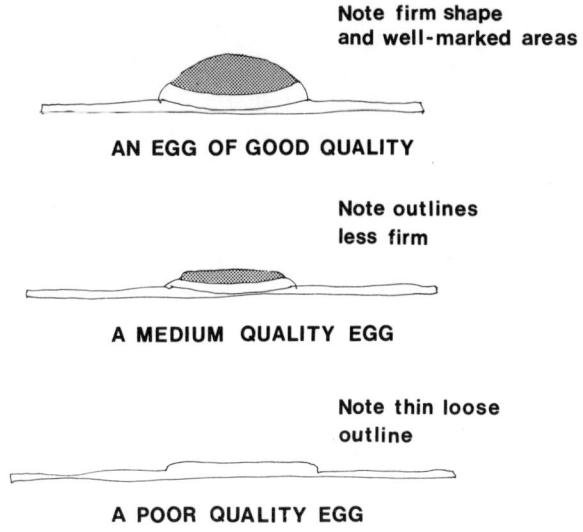

Fig. 14

Note firm shape
and well-marked areas

AN EGG OF GOOD QUALITY

Note outlines
less firm

A MEDIUM QUALITY EGG

Note thin loose
outline

A POOR QUALITY EGG

(a) *Soft-shelled eggs*. These are the most common faulty eggs and they usually occur near the start of lay while the birds are settling into a routine. Sometimes a number of shrivelled skins will be found on the droppings board below perches. This is also normal during the period of adjustment. Should the soft shells continue for a number of weeks consult your feed supplier or merchant.

Occasionally, soft-shelled eggs will occur towards the end of lay, and prior to a moult, along with extra-small eggs. In fact, the normal pattern of egg production is a high proportion of small eggs in the first few weeks, with some soft shells. As numbers build up around thirty weeks of age so does the egg size, putting most of them in the Standard weight range. As the year progresses the number of eggs decline but the proportion of Large increases. Egg size may well tail off again at the end of lay.

(b) *Cracks*. Cracked eggs are usually caused by the shell not being sound when it is produced or through management failures such as mishandling on collection, insufficient soft material in the nest box, or the birds being disturbed in lay. Birds rarely lay cracked eggs—the damage invariably occurs after. There is just a possibility that the shell may be particularly thin and susceptible to damage. If it persists have another word with your feed supplier or merchant.

(c) *Mis-shapen eggs*. Ailments like Newcastle disease and infectious bronchitis, which affect the reproductive system, can result in mis-shapen eggs as well as thin shells. Often, older birds near the end of lay will produce eggs with wrinkles running from end to end. You may find that the internal quality is poor as well, but there is little you can do about it. It's a sign of old age.

Flat-sided eggs are most common in young layers and result from two eggs in the shell gland at once. Often a bird will lay one of these and a normal egg in the same day. There is nothing wrong with the eggs and the condition will usually clear up, although in a few instances the habit sticks with the bird for life.

The same story applies to eggs with a bulge of raised shell at the narrow end. There is not much you can do about it but then it does not do any harm. The egg will still taste delicious.

If the egg shell looks rather like the surface of the moon—i.e. very rough to the touch—it is a fair indication that your birds are old or diseased—more likely the former. The condition is caused by loss of muscle tone in the shell gland about which you can do nothing, but again it is harmless.

(d) *Dirty shells*. Do not be alarmed if egg shells carry smears of blood. This is normal with young layers providing it is not excessive and does not persist for more than a few weeks. If the shells are particularly dirty take a look at the nest box. Clean out any droppings and cracked or broken eggs, change the litter regularly and clear out the dust. Remember that when the egg is laid it is warm and moist and picks up any dirt with which it comes into contact.

Internal quality
The time for assessing the internal quality of the eggs will be when a few are broken out for frying or whisking.

(a) *Runny whites.* If the white runs all over the pan, it is likely that the egg is an old one or that it has been stored at too high a temperature. Alternatively, the egg may be particularly fresh, i.e. under 24 hours old.

The birds may be suffering from one of the respiratory diseases mentioned earlier or the feeding may be wrong and the birds are short of protein. Both possibilities are unlikely and are difficult to prove unless the disease is particularly obvious for other reasons.

In a first quality fresh egg, the yolk stands up well and is held centrally by the albumen. Both chalazae (a pair of cords which hold the yolk in the centre of the albumen) are distinct and firmly attached to the yolk. The yolk appears firm and rotund and with no signs of mottling. Yolk colour is an even yellow; the albumen is clear with no tint or colour and the outer thick and thin albumen is distinct. Usually the germ cell is just visible on the yolk as a small disc. This is quite normal.

(b) *Pale yolks.* If the yolk appears unusually pale this suggests insufficient colouring pigment in the diet. More maize or grassmeal should clear this up—try adding some grass cuttings to the feed or feeding fresh greens.

(c) *Blood and meat spots.* The egg contents may be marred by small blood or meat spots in the yolk or albumen. Blood spots come from minor haemorrhages as the yolk is formed or as it passes along the oviduct. Meat spots are either small pieces of tissue from the oviduct wall or small pieces of egg material fragmented during formation.

Although both these inclusions are a little unsightly they are harmless and will usually rectify themselves in a short time.

Egg candling
Commercial egg producers or egg-packing stations selling eggs to the public are obliged to candle the eggs. It is the best available method of assessing internal quality on a large scale without actually breaking the shells. The process involves viewing the egg against a strong light which penetrates the shell and outlines the contents. Virtually all egg candling is done on a mass scale on large, sophisticated machines but the small producer who takes an interest in this side of things can still purchase single egg candlers. Or better still he can save himself some money and construct a perfectly adequate device as shown in Fig. 15.

The source of illumination is a 60 watt clear glass electric light bulb. The cushion of felt around the aperture reduces leakage of light when an egg is firmly applied to it. The diameter of the hole should not exceed $1\frac{1}{4}$ in. To maximise efficiency, the candling area should be as dark as possible.

The egg is held, broad-end uppermost, at a slight angle to the aperture of the lamp and is twirled on its axis. The sort of things to look for are shown in Fig. 16.

In a fresh egg the air space is about $\frac{1}{4}$ in. deep and stationary when the egg is twirled. The yolk will appear as a shadow with a distinct outline. It will not move appreciably away from the centre when the egg is moved. In older eggs, the air space is larger.

82

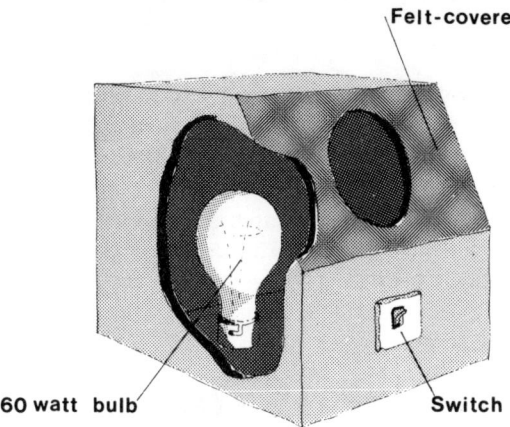

Felt-covered top

60 watt bulb

Switch

Fig. 15 A home-made egg candler. Anyone wishing to take egg quality checks a step further can construct a single egg candler. All that is needed is a 60 watt bulb, a bulb holder connected to the electricity supply, a switch and some spare pieces of plywood. Make the candler about 10 in. × 10 in. with a 1 in. hole on the angled side and a felt covering around the hole. Switch on and take a closer look inside the egg by holding it broad end uppermost at a slight angle to the aperture, and twirl it on its axis.

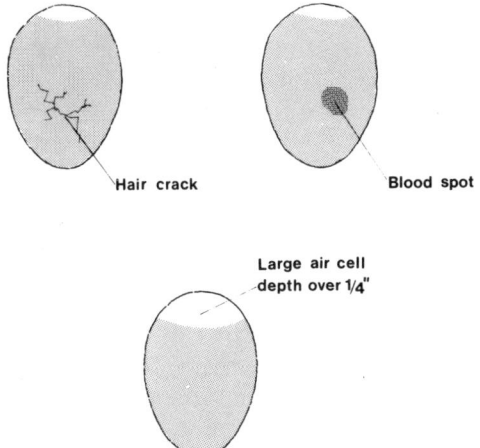

Hair crack

Blood spot

Large air cell depth over 1/4"

Fig. 16 Egg faults. These are some of the faults that will show up if you use an egg candler. None of them are harmful and the eggs will still taste good, but anyone selling a quality product would have to remove them. The hair crack could well result from rough handling; the blood spot will occur from time to time; and the enlarged air cell indicates that the egg is not particularly fresh.

Foreign bodies such as meat and blood spots will become apparent so if you are selling eggs to the public it would be wise to remove these specimens. Hair cracks can also be seen. The first step in establishing a quality image for your produce is ensuring that none of these faulty eggs are passed on to customers.

8 The End of the Road

It is an unfortunate fact that layers eventually come to the end of their useful productive life after one or possibly two laying years. If left alone they would probably live to a ripe old age but long before that they will represent a hefty feed bill, give very erratic egg production and be just a sleeping partner in your operation. Replacement is the only answer, just as you must pull out the non-productive first-year birds if you take the rest on to a second year.

Having made the big decision to replace, the next step is to convince yourself that each one of the culls represents a potential weekend joint. Anybody who has ever kept a few chickens in the back garden knows how agonising it can be to think of eating your feathered friends. One way out of the quandary is to advertise them as live eating birds.

But really this is no answer, for the price will be relatively low and anyway the end product is just as much a part of the home produce as the eggs. Although they are termed 'old hens' they are still comparative youngsters and can be turned into a range of extremely tasty dishes with little effort.

Killing the bird

If you can't do it yourself call on a friend or neighbour who is a bit more detached. The commonest method of killing a small number of birds is by dislocation of the neck. Isolate them from the feed and the run 24 hours beforehand. Then, holding the legs and tips of the wings in the left hand (assuming you are normally right-handed) with the bird hanging head downwards, place the first and second fingers of the right hand on either side of the neck with the palm of the hand facing away from you so that the neck is in a fork. Use the other fingers to press the head and comb back into the palm of the hand.

Carry out this operation in the standing position and direct the bird diagonally across the front of the right thigh. Using firm steady downward pressure, so that the fork formed by the fingers is pressing downwards, dislocate the neck. An easing of the resistance will indicate when the neck has gone. Be sure not to stretch too hard at this point or the head will be severed completely. A cavity about 2 in. long will be formed by this action and at the same time the main arteries in the neck will be severed. The blood drains into this cavity. Death will be instantaneous.

As soon as the neck is dislocated, the body will start to flutter quite violently. This is a totally insensitive nervous reaction to the severing of the spinal cord. It only lasts for a short while but can be a little alarming. Keep a firm hold on the legs and head until it passes. If the fluttering is unacceptable, have a loop of string hanging in the garage or from the branch of a tree ready to tie round the feet and leave the bird to hang for no more than a minute before the next stage.

Plucking

Feather plucking is a messy job so be warned. Wear your oldest clothes and have an old bin ready to put the feathers in. The best place to carry out the operation is in an old shed or the garage where there are no strong draughts to blow the feathers about. Sweeping up afterwards will then be easier than chasing feathers all over your garden and, possibly, the garden next door.

The easiest position in which to pluck is sitting down, so hang the bird by its feet at a convenient height. Choose a different hanging height if you prefer to stand up.

Start plucking as soon as possible after killing, certainly within two minutes, for the feathers come away far easier when the carcase is still warm. Remove the flight and main tail feathers first by three quick pulls. The breast feathers should be removed next as these come out easiest from a warm bird. There is a feather tract on either side of the breast and special care must be taken in plucking to avoid tears in the skin. These not only spoil the appearance, but can impair the keeping quality if the birds are not to be cooked straight away. To avoid this hold the skin with one hand, and with the other hand pull out small groups of feathers in the direction in which they lie. This should be repeated over the entire body. On completion of this stage the bird is said to be rough plucked.

Next comes the stubbing where short feather stubs are removed with a blunt knife and the thumb which combine to act like tweezers. When all the feathers and all the stubs have been removed there is still likely to be a mass of hairs over the carcase. These can be removed by holding the bird over a gas or methylated spirits flame—the latter being preferable and easy to produce. Just put some meths into a tobacco tin and apply a match. Finally, the vent should be squeezed to get rid of any excreta and the bird's feet should be washed.

Alternative plucking method

If more than one or two birds are being done at the same time, they can be plucked by the soft scald method. The carcase is immersed in water at a temperature between 124°F and 127°F for between 30 and 120 seconds. The feathers can then be removed easily and, providing the temperature range is not exceeded, the carcase appearance may even be improved.

Shaping the carcase

As it cools, the carcase will start to stiffen up and you can take advantage of this process to shape the bird. A piece of string 9 in. long is tied to the middle toe of each foot, the hocks pushed back into the body, and the strings taken down along the thighs and tied across the back of the bird. A 12 in. piece of string is then passed in front of the hocks, crossed behind them and tied around the parson's nose. This throws the breast into prominence.

The birds should be kept tied up like this for at least 24 hours in a cool, or preferably cold place. A cellar or the bottom of the refrigerator are ideal places.

Trussing

This is the process whereby the innards are removed. The alternative terms are gutting or evisceration. It is the messy end of the operation. Equip yourself with a decent sized chopping board, some string and a sharp knife. All the better if you can work on a surface that can be scrubbed clean afterwards.

First remove the strings that were used to shape the bird. Next remove the sinews in the legs. This is done by making a light cut around each shank about an inch below the hock joint. Break through the bone, by bending the foot over the edge of the table, and withdraw the thick white sinews by steadily pulling on each foot. It helps to hold the chicken leg with a cloth to avoid unnecessary abrasion. You may need to loosen the tissues with a skewer.

Next comes the gutting stage. First job is to remove the head. Place the bird breast downwards and make a cut across the back of the neck at a point about 1 in. above the shoulder. This cut should go a little less than half-way across the neck, leaving a larger flap of skin at the front than at the back. The skin is then cut down almost to the shoulder line and then up towards the head. When the flap of skin at the front is about 3 in. long the excess skin may be cut off. Then, holding the neck and head in the left hand, either twist this off or cut through the joint nearest the body. The flap of skin left will later be used to close the opening left behind.

The crop now needs removing. Insert a finger into the neck cavity and work the crop round until enough is exposed to make a cut low down in the neck cavity. The crop can then be cut away from the gizzard and the rest of the intestines. The wind pipe can be removed at the same time.

The bird should then be turned onto its back and a finger again inserted into the neck cavity to loosen the lungs and other organs. This makes their subsequent removal simpler, but on no account should they be taken out at this stage. Next, stand the carcase on the neck end and make a transverse cut between the vent and the tail. Insert a finger through this cut into the body. Loop the finger round the underside of the vent. This will help to define the outline of the vent which can then be completely cut away from the body with the intestines still attached to it.

With the bird again on its back, enlarge the initial cut into the abdomen by slitting a little way towards the breast bone. It is possible that substantial amounts of fat will be found in the body at this stage. Remove this, then draw out the rest of the contents by taking hold of the gizzard and drawing this organ together with the intestines, liver, heart, lungs, etc. out through the tail end.

Retain the gizzard, heart, liver and neck for cooking as you see fit. But first skin the gizzard by splitting it lengthwise with a knife and removing the inner horny skin that holds food material, and detach the small, green gall bladder from the liver with a sharp knife, taking care not to puncture it.

The bird is now ready for tying-up. Washing the carcase at this stage is not normally necessary, but if the intestines or the gall bladder have been punctured, for instance, then a quick rinse with cold water may be necessary. Otherwise wipe over with a damp cloth—inside and out. At all times ensure that knives, hands and surfaces are thoroughly washed and do not come into contact with any other foodstuff or food area until they are clean.

The carcase can either be tied up, which will make it more attractive for eating or storing as well as more manageable for cooking, or cut into portions. On balance it is probably worth tying the bird up properly so that it can be cooked whole—or stored in the freezer for a time—and then jointed for casseroling later. It is unlikely that a roasted old hen will be tender enough to provide a suitable product for serving up straight, although a young, plump first-year bird can be dressed up sufficiently to make an acceptable roast.

Tying

There are two useful methods of tying the bird. The first, a simple method, is perfectly adequate. The second is more professional.

With the bird on its back take 18 in. of string and place it under the back of the bird, bringing it up the sides of the body under the wings. Still holding both ends of the string, bring them over and then under the legs forcing them as far forward as possible in order to push out the breast flesh. Now take them back to the front of the bird, under the legs and over the wings as close as possible to the body. Turn the bird over and tie the strings tightly across the back.

The wings should now be turned under the bird with another piece of string. Tie the two legs together at the knuckle joints and tie down under the parson's nose. Finally, push the breast flesh upwards from the sides of the bird, drawing the skin well forward and then tucking it under the wings.

The alternative method, shown in Fig. 17, makes use of a 10 in. trussing needle. Thread the needle with 15 in. of string, place the bird on its back and fold the flap of skin over the neck opening. Fold the wings over the back by turning the ends under the body. This will keep the flap in place. Pass the needle through the wing below the 'elbow', under the bird's back and through the other wing in the corresponding position. Then press the legs forward and down and push the needle through the bird behind the legs and low down in the thigh to come out at the same point on the other side. Tie the two ends of the string at the side of the body.

Re-thread the needle and with the bird still on its back, pass it through the flesh under the pelvic bones, over one shank, through the flesh just below the end of the breast bone and over the other shank. The needle is then withdrawn and the ends of the string are tied tightly on one side of the body.

To cook or freeze?
The bird is now ready either for cooking or freezing. The chances are that if you have killed more than one bird in a day, freezing will be the answer. As already

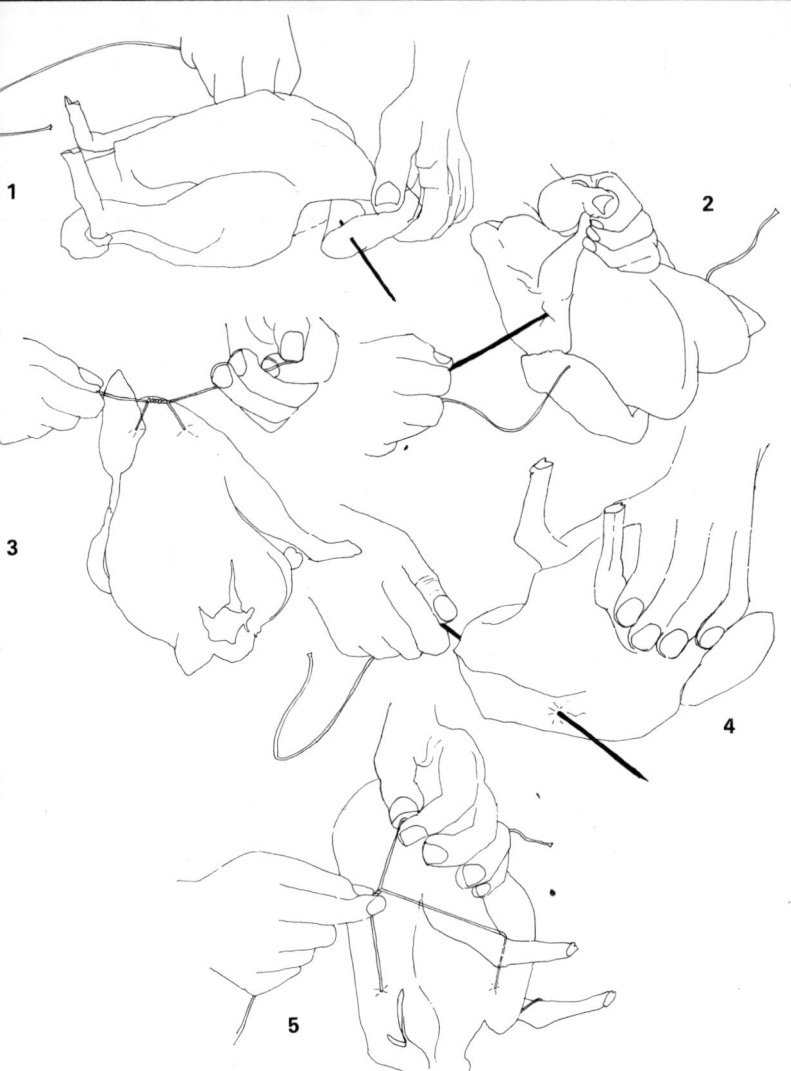

Fig. 17 Five easy steps in tying a bird.
1. After threading the trussing needle, fold the wings over the back and pass the needle through them at the points shown.
2. Press the legs downwards and pass the needle through the body from thigh to thigh.
3. Bring the two ends of the string over the back and tie, making sure that the flap of skin is covering the neck hole.
4. Pass the needle through the skin under the pelvic bones as shown.
5. Take the string round one hock joint, through the skin at the tip of the breast bone, over the other hock joint and tie.

mentioned it is unlikely that you will want to serve the birds as straight roasts since they tend to be a little tough, flavourless and short of meat. But the following recipe has been used to good effect with many an old hen:

> Rub the breast with lemon juice. Cut off the legs to the first knuckle. Place enough water in a pan to cover the bird. Add 2 medium-sized carrots and a sliced onion. Season and add a bouquet garni. Bring the water to the boil, add the chicken and fast boil for 10 minutes. Simmer gently for another 2–3 hours until the leg meat is tender. Lift out, drain and serve with a savoury white sauce or bread sauce.

Otherwise, boil the chickens in an ordinary pan or pressure cooker and prepare some chicken pies, casseroles, curries and the like, and freeze them.

One very important thing to re-emphasise is that cleanliness is paramount when you are gutting and trussing your own birds in this way. Not only must you avoid contaminating the carcase but you must ensure that nothing else around is handled with wet chickeny hands.

9 The Chicken and the Egg

The modern layer is a machine which takes in food and uses it to produce her own running power, to carry out repairs and to make eggs. Obviously, as you are going to be in charge of such a machine it helps to know something about its parts and how they work, particularly those connected with the feed that you put in and the eggs it puts out.

Where the food goes

Figure 18 highlights and identifies those external features of chickens that are useful to know. It also shows the digestive system and some other relevant organs.

When a beakful of food is eaten and swallowed it passes first into the crop, which is, in effect, a store where a quantity of food can be kept until it is digested. This enables the bird to swallow a meal fairly quickly and retire to a safe place where it can be digested. The crop also softens the food.

Next, the meal passes into the stomach or pro-ventriculus, where the digestive juices are added, before passing into the thick muscular gizzard. Here large particles of food are ground up so that the digestive juices can do their work. The grit that the hen eats collects in the gizzard and takes the place of teeth in aiding the grinding process.

Fig. 18

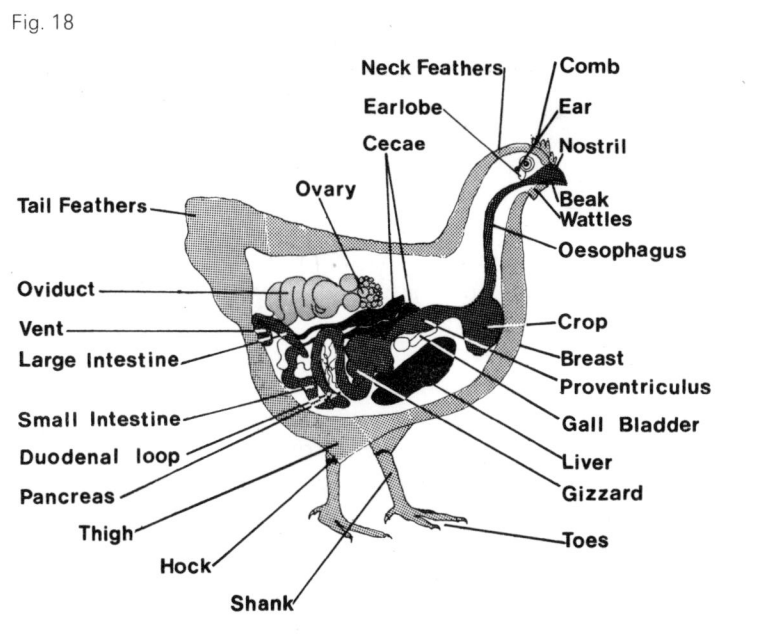

Absorption of the digested food into the blood stream takes place along the length of the intestine which ends up as the short, large intestine where water is extracted from any waste food not absorbed. This waste then passes on to the vent where it is expelled as droppings.

How the eggs are made

An egg is formed by the reproductive system of the hen, the ovary and the oviduct being the two parts most closely involved. Each egg starts as a minute cell (ovum) in the ovary. Attached to each is a supply of food, for its natural function is to become the ideal environment for a fertilized cell to develop into an embryo which will survive in the shell outside the bird's body for about twenty-one days without additional food. This only takes place if the egg is fertilized.

Although the eggs with which we are concerned are unfertilized—unless we are breeding—the process still continues in the same way. Eggs are still produced at the same rate whether a cockerel is there to mate with the hen or not.

After about seven to ten days this developing yolk, as it is called, is ovulated. It becomes detached from the ovary and enters the first part of the tube-like oviduct, which is about 30 in. long. As the yolk progresses along the oviduct the white (albumen), the shell membranes and the shell are all added in sequence until the egg is laid, about 24 hours after leaving the ovary.

The double-yolked egg is caused by two yolks being released in close succession and ending up in the same shell. If no shell is produced it is probably a lack of calcium, and if the shape is uneven it is probably due to a temporary fault in the muscles of the oviduct.

Design of the egg

From Fig. 19 we can see that even the unfertilized egg which we eat has a fairly complicated structure. At the centre is the yolk which is uniform yellow except for the white germ cell which is the part that would develop into a chick had the egg been fertilized. The yolk is held centrally, away from the shell, by the chalazae which are fibrous portions of the white. It is cushioned by a thin and a thick layer of white.

On storage, the chalazae tend to become detached and the white becomes runny so that the yolk may well fall to one side of the shell. The runny white is a characteristic sign of an old egg just as thick white and distinctive chalazae are signs of a fresh one (see Chapter 7 on egg quality).

The air space at the thick end forms when the warm egg contracts as it cools after being laid. One of the standard ways to test the freshness of the egg is to measure the size of the air space—the larger it is the older the egg.

On the inner surface of the shell are two cell membranes. Then comes the calcium carbonate shell, which contains numerous minute pores as well as

Fig. 19 Structure of the egg.

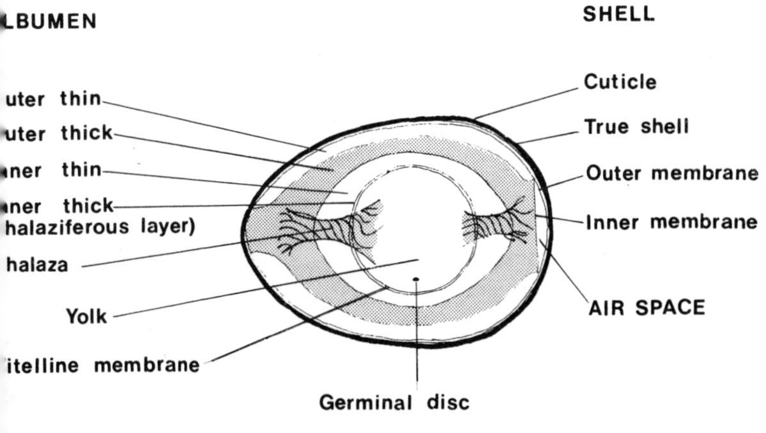

ALBUMEN

uter thin
uter thick
ner thin
ner thick
(chalaziferous layer)
halaza
Yolk
itelline membrane
Germinal disc

SHELL

Cuticle
True shell
Outer membrane
Inner membrane
AIR SPACE

colour pigments. Finally, there is a thin transparent cuticle on the outside. This gives the characteristic bloom to a fresh egg—again it disintegrates on storage and its presence or otherwise gives a guide to age.

Why not an egg a day, every day?

If hens laid an egg every day we would, quite naturally, expect to get 365 eggs from each bird in a year. But generally, we get between 240 and 300 which is only 68–82 per cent of the theoretical maximum.

To understand why this is, we have to know something about the laying mechanism. Going back to the formation of the egg, it was mentioned that yolks are released into the oviduct at the rate of about one a day.

Strictly speaking, we should talk in terms of hours. Indeed, the best layers do ovulate once every 24–25 hours, and a new yolk is released about half an hour after the previous egg is laid. So we get an egg a day or more, although each one is laid about half an hour later in the day than its predecessor.

But even with the best layers, this does not go on indefinitely. A hen may lay for twenty-four days in a row but then takes a rest, possibly only a day, and then lay for another twenty-four days on the trot. The eggs laid in these periods are collectively called a 'clutch' and the periods between are 'pauses'. It is these clutches and pauses that cause eggs to be lost from the theoretical total of 365 eggs a year.

In the less prolific layers of a flock—even six birds can be considered a flock and will be made up of individuals—these clutches will be even more defined. There will be fewer eggs to a clutch, perhaps only three, the pause between clutches will be longer and the time between ovulations may be as long as 28 hours.

Other reasons for falling short of the 365 goal, apart from bad feeding, disease or poor lighting, are that in the first few weeks of lay, laying is irregular with long pauses as the reproductive system settles down. Eggs tend to be small at this stage and some have soft shells.

As the laying year progresses the reproductive system settles down. Losses in production during the mid-lay period are mainly due only to the pause days when no ova are released from the ovary. Continuing on into the year the hen's reproductive activity declines and egg numbers tail off, although the eggs tend to be larger.

The effect of light

Light, natural or artificial, plays a vital role in egg production. More correctly, it is the day-length that is the influence. The mechanism is complex but suffice to say that it affects a small organ behind the eye which in turn sends messages to the ovary and affects the ovulation process.

If chickens were left to their own devices they would start to lay in the spring, stimulated by the increasing day-length. They would lay a lot of eggs over a short time and then rapidly go out of lay as the shorter days of autumn and winter approach. Finally, they would go into an early moult. In effect they burn themselves out early because they come into lay early, long before they are mature.

The professional egg farmers hold back birds from laying until they are at least twenty weeks old, by the use of artificial light. Rearers ensure that the day-length the birds receive is never allowed to increase in this period. So that if, for instance, chicks were hatched in March or April when natural day-length is increasing, they would be kept indoors in a house where the only source of light is electric and this would be kept constant at about eight hours of light a day and sixteen hours of darkness.

The golden rule with a laying bird is never allow day-length to decrease. What we do, in fact, is to provide artificial light to supplement the natural light and bring it up to spring and summer levels of fourteen to seventeen hours a day (see Chapter 5). The important thing is to ensure that once a level has been set, it is not varied. Providing artificial light in this way will not produce more eggs; it will even out production so that you are not faced with a spring flush followed by a protracted shortage of eggs in winter.

10 Brooding and Rearing

Keeping chickens, like any interest worth its salt, will get a hold on you. Despite initial assurances to friends that you are only in the job for the eggs you get out of it, that you are not interested in rearing your own chicks, and that you prefer to buy your pullets ready-grown, the time will come when you start looking at the possibilities of raising your own stock. But there is more to it. It is like the gardener who starts by buying boxes of plants and then realises it is more fun to buy packets of seeds and do it himself.

It is fun, without a doubt. There is nothing to beat the satisfaction of knowing that you have raised your adult laying flock from day-old. But do not delude yourself into thinking it is cheaper. Pullets are very competitively priced these days, and if you cost the extra hours you are obviously going to spend getting your chicks to point-of-lay you certainly will not be in profit.

It goes almost without saying that, as with pullets, you should go to a reliable source for your chicks. Do not be surprised if you get one or two chicks more than you ordered. Reputable breeders guarantee safe arrival and will send replacements for any dead ones so long as you return them straight away.

The same guarantee covers sickly looking birds. Healthy chicks will be trying to jump out of the box as soon as you remove the lid. So take a careful look at the stragglers who are not so keen. Unless you collect the chicks yourself, the hatchery will deliver or send them by British Rail.

Off to a good start

The initial brooding can be by broody hen or by using an electric or gas brooder. But whatever method you choose have everything ready for action well before the chicks are due to arrive.

If you have a broody waiting, make sure she has got accustomed to the prospect of coping with a clutch of chicks and have her sitting on some eggs (china eggs can be bought for the purpose) for at least a week beforehand. Hens have a reputation for being stupid, but they are not so daft as to think that they can acquire a family of chicks without first sitting on the eggs.

Leave the chicks in the box when they arrive and introduce them in the evening unless they seem particularly hungry in which case do it at any time.

Introductions should be conducted with care. Take just one chick from the box and put it carefully under the hen. Do not move her, but shut the nest and leave her with the new arrival for an hour. To check how well the chick has been received, place it before the hen and scatter a few grains of corn. If the hen invites the new arrival to share the snack, she is accepted. You can introduce the rest of the family, but only a few at a time, spreading the introductions over three or four hours, depending on the number of chicks. Again it is all part of the deception. No hen will believe she can hatch a dozen chicks all in one go. Once she has taken to them, it will be safe to move the family to the quarters in which it will grow up.

If the initial introduction to the first chick does not go too well, give the broody another half an hour or so and repeat the process. In the meantime, the remaining ones in the box will be getting peckish so give them a meal of chick starter crumbs and a drink and put them back.

There are two basic types of artificial brooder—the infra-red lamp and the electric hen. The infra-red is ideal for small numbers of chicks and a 250 watt lamp will be powerful enough for up to fifty chicks. More than this and you will need two lamps set above the birds 18 in. apart.

Temperature can be controlled by adjusting the height of the brooder over the litter with 98–100°F as the target. A red or blue spirit thermometer will give a more accurate reading than the ordinary mercury type.

Fig. 20 An electric hen brooder. According to size, electric hens will take up to 200 chicks. They act like an overhead electric blanket and, like the chick lamps, are raised as the birds get older.

Place the thermometer on the litter, about 6 in. from the centre of the lamp and adjust the height of the lamp until you have a reading of 100°F. As the chicks grow, the lamp can be raised and the heat reduced.

Infra-red lamps need no curtains or canopies and the chicks are visible all the time. Normally, they will spread out at night and sleep comfortably in the warmth of the lamp. If they huddle together, it is a sign that the heat is not getting through so temperatures should be checked.

One possible trouble spot will be the reflector which must be kept bright and clean for a dull emitter type of lamp. With the bright emitter type, the reflector is built into the lamp where it cannot get dirty.

The electric hen brooder operates like an electric blanket suspended above the chicks (see Fig. 20). The legs are adjustable so that the 'blanket' can be raised as the chicks grow. Manufacturers claim that one unit will take up to 100 chicks and although the initial cost (in the region of £25) is considerably more than that of an infra-red lamp, the brooder is cheaper to run. One disadvantage is that it is necessary to stand on one's head to see the chicks under the brooder, but the unit works very well, nevertheless, and the birds normally thrive.

Whatever type of brooder you use, buy the best one available in that type. A good one will last many brooding seasons and will be less likely to break down just when you need it most. But however reliable it has proved, always check your equipment well before the new batch of chicks is due and have it switched on and warmed up ready to give them the right reception.

Chicks will need little instruction on how to use the brooder and their greatest need in the first couple of

Fig. 21 In the early days, a cardboard surround will confine the chicks close to the lamp and prevent them from getting lost. As they age, the level of the lamp can be raised and the area inside the surround increased.

days will be warmth. After the initial 48 hours they will also need food, but the order of priorities remains the same—warmth and then grub. Do not worry about the outside temperature. Your young flock will survive quite happily in temperatures below freezing so long as there is somewhere warm to come home to.

Chicks need to be confined to within about 12 in. of the edge of the electric hen brooder (see Fig. 20). This means a low fence, no more than 18 in. high, of cardboard, hardboard, or similar material. Cardboard cut from the boxes in which the groceries arrive can be used. It can be joined by clothes pegs.

The surround for infra-red lamps should have a diameter of about 3 ft for the first fifty chicks. Beyond this total, add an extra foot for every additional fifty.

Place food and water in the area between the edge of the brooder and the surround. Never provide meals under the brooder. If the chicks do not come out to feed the brooder temperature is not high enough.

Ideally, chicks should sleep just outside the area

covered by the brooder. If they crowd against the surround after a couple of weeks, enlarge the area. At three weeks, they should have the run of the brooder house unless it is very large.

In the first stages, check that the chicks return to the warmth of the brooder after a meal. If they have too much room in which to move around they will tend to lose their way and huddle into groups. This will also happen if the brooder becomes too cold.

Dry sand is an ideal litter for the first few days, followed by peat moss or shavings. All the material should be perfectly dry and mould-free.

If you use a broody hen to raise the chicks they will need to be housed in a coop with a run. Rats are very partial to young chicks and to keep them at bay the coop must be capable of being shut up at night. It should have a solid floor under the nest box area.

First eating steps

The development of proprietary baby chick crumbs, mashes and crumbles, has taken much of the mystique out of feeding young stock. The crumbs and crumbles tend to cost more than the straight mashes and are more palatable but, except where the chicks are being reared under a broody hen, are not really necessary.

Chick crumbs can be made available to broody reared stock in a shallow trough, while the 'mother' hen will be satisfied with 4 oz of whole wheat a day. She may sample the crumbs, and it will not harm the chicks if they try her wheat. The broody will stay with the chicks until she comes back into lay after about five weeks.

Birds reared by artificial means will be quite happy with a proprietary mash which can be fed for the first few days on sheets of paper in a shallow tin lid or in Keyes trays—the fibre pulp trays which hold thirty eggs. Proper troughs, low, open with or without a spinner, can be put down after two or three days. There are special chick troughs available with a grid on top through which the chicks are fed, but the birds soon outgrow them.

Trough space requirements in the first sixteen weeks are:

0 to 3 weeks	2 in.
4 to 8 weeks	3 in.
9 to 15 weeks	4 in.
16 weeks plus	4–5 in.

If the trough is double sided count both sides. A 24 in. double-sided model will have a total space of 48 in.—enough for 16 chicks from 4 to 8 weeks. The plastic lid of a date box can be used as a trough for the first two weeks. It is about 9 in. long and two sided which allows space for a dozen birds.

Shallow baking tins can also be pressed into service and after two weeks a growers' trough can be used. A spinner running the length of the trough will keep the chicks' feet out of the mash. Do not overfill the trough—halfway up the side is ample.

Drinking tackle

The best and simplest way to give chicks a drink is with an upturned jam-jar fitted to a patented base (see Fig. 22). The water is fed into a shallow trough. These jam-jar drinkers will carry the birds through the first four weeks after which normal adult drinkers can be used. Don't forget—water is vital to a chick and it should never be without it. The following is the daily water requirement for each twenty-four birds:

0 to 2 weeks	$1\frac{1}{2}$ pints
3 to 5 weeks	5 pints
6 to 10 weeks	8 pints
11 to 18 weeks	10 pints

Fig. 22 One of the simplest but most effective chick drinkers is an up-ended jam-jar fitted to a patented base which acts as a trough. Top: this ½ gallon drinker will serve 25 birds and is ideal for growing stock.

Feed chick mash for the first six weeks, adding quality scraps (meat, fish or milk pudding left-overs) from three weeks onwards. At seven weeks introduce a proprietary growers' mash with a coccidiostat until the fourteenth week of age. The coccidiostat is a drug mixed in the ration by the feed compounder to prevent the development of coccidiosis—a disease caused by the coccidia eggs which can exist in the litter. At fifteen weeks switch to a growers' mash which does not have the coccidiostat.

The grit box can be introduced at seven weeks, at the same time as the growers' mash, and fed from then onwards.

Appendix A

Useful addresses

Help and Advice
Animal Aids Ltd,
17 Stratford Road,
Salisbury, Wilts.
Tel: Salisbury 29663
(Stockists and suppliers of
poultry medicines, vaccines,
disinfectants. All supported
by a laboratory diagnosis and
advisory service.)

Animal Medics Ltd,
Elliott House,
Greenacre Road,
Oldham Lancs. OL4 1HB
Tel: 061-652 1307
(Stockists and suppliers of
poultry medicines, vaccines,
disinfectants and equipment.)

British Egg Information
Service,
37 Panton Street,
London SW1 4EW
Tel: 01-839 7258
(Egg recipe leaflets, egg cookery
books and egg gadgets.)

Eastern Counties Farmers Ltd,
PO Box 34, 86 Princes Street,
Ipswich, Suffolk IP1 1RU
Tel: Ipswich 56071
(Suppliers of animal health
products in East Anglia.
Independent post mortem
service.)

Salisbury Laboratory Ltd,
17 Stratford Road,
Salisbury, Wilts.
Tel: Salisbury 29663

Poultry World,
Surrey House,
1 Throwley Way,
Sutton, Surrey
Tel: 01-643 8040
(Advisory service from only
weekly journal in the poultry
industry.)

The Poultry Club,
Virginia Cottage,
6 Cambridge Road, Walton
on Thames, Surrey
(Details of all the breed
societies.)

Ministry of Agriculture,
Regional Advisory Offices.
(Advice on all aspects of
poultry keeping.)

North:
Government Buildings,
Kenton Bar,
Newcastle-upon-Tyne
NE1 2YA
Tel: 0632-86 9811

Yorkshire and Lancashire
Region:
Block 2, Government
Buildings, Lawnswood,
Leeds LS16 5PY
Tel: 0532-67 4411

East Midland Region:
Shardlow Hall, Shardlow,
Derby DE7 2GN
Tel: 0332-792-313

West Midland Region:
Woodthorne,
Wolverhampton,
W. Midlands WV6 8TQ
Tel: 0902-754190

Eastern Region:
Block C,
Government Buildings,
Brooklands Ave., Cambridge
CB2 2DR
Tel: 0223-58911

South Eastern Region:
Block A, Government Offices,
Coley Park, Reading, Berks.
RG1 6DT
Tel: 0734-581222

South Western Region:
Block 3,
Government Buildings,
Burghill Road, Westbury-on-Trym,
Bristol, Avon BS10 6NJ
Tel: 0272-500000

Wales:
Trawscoed, Aberystwyth,
Dyfed
Tel: 0974-3255

Ministry of Agriculture,
Veterinary Investigation
Centres
(Birds for post mortem
examination at a charge of
£6.60 for a single bird or
£8.90 for batches.)

Aberystwyth: Y. Buarth,
Aberystwyth, Dyfed, Wales
SY23 1ND
Tel: 0970-2374

Ashford: Coldharbour, Wye,
Ashford, Kent
Tel: 0233-81 2315

Bangor: Penrhos Road, Bryn
Adda, Bangor, Gwynedd,
Wales
Tel: 0248-2561

Bristol: Langford House,
Langford,
Bristol, Avon BS18 7DX
Tel: 093 485 2421

Cambridge: Field
Laboratories,
Madingley Road,
Cambridge CB3 0ER
Tel: 0223-58691

Cardiff: Government
Buildings,
66 Ty Glas Road, Llanishe,
Cardiff, S. Glam., Wales
CF4 5ZB
Tel: 0222-757971

Carmarthen:
Job's Well Lane,
Johnstown, Carmarthen,
Dyfed, Wales
Tel: 0267-5244

Exeter: Staplake Mount,
Starcross, Exeter, Devon
EX6 8PE
Tel: 062-689 481

Gloucester: Elmbridge Court,
Cheltenham Road, Glos.
GL3 1AP
Tel: 0452-21421

Leeds: Quarry Dene,
Weetwood Lane, Leeds
LS16 8HQ
Tel: 0532-52636

Lincoln: Riseholme, Lincoln,
Lincs.
Tel: 0522-31201

Liverpool: University of
Liverpool, Vine Street,
Liverpool L7 7EH
Tel: 051-709 5848

Loughborough: The Elms,
College Road, Sutton
Bonington, Loughborough,
Leics.
Tel: 0509-72332

Newcastle-upon-Tyne:
Whitley Road, Longbenton,
Newcastle NE12 9SE
Tel: 0632-66 2292

Northampton: Pitsford Road,
Moulton, Northampton
NN3 1RS
Tel: 0604-45781

Norwich: Jupiter Road,
Norwich, Norfolk NOR 9ON
Tel: 0603-46278

Penrith: Merrythough,
Calthwaite, Penrith, Cumbria
CA11 9RR
Tel: Calthwaite 85295

Reading: Block A,
Government Offices, Coley
Park, Reading, Berks.
RG1 6DT
Tel: 0734-58 1222

Thirsk: West House, Station
Road, Thirsk, N. Yorks.
YO7 1PZ
Tel: Thirsk 22065

Truro: Polwhele, Truro,
Cornwall TR4 9AD
Tel: 0872-2150

Weybridge: Weybridge
Centre, Woodham Lane, New
Haw, Weybridge, Surrey
KT15 3NB
Tel: 093-23 4111

Winchester: Itchen Abbas,
Winchester, Hants.
Tel: 0962 69411

Wolverhampton:
Woodthorne,
Wolverhampton,
W. Midlands EV6 8TQ
Tel: 0902-754190

Worcester: Block C,
Government Buildings,
Whittington Road, Worcester,
Worcs. WR5 2LO
Tel: 0905 353514

*Dept. of Agriculture for
Scotland, Poultry Advisers*

West of Scotland Agricultural
College, South Western
Agricultural College Office,
Edinburgh Road,
Stranraer
Tel: 0776 2649

East of Scotland College of
Agriculture, West Mains
Road, Edinburgh EH9 3JG
Tel: 031-667 1041

North of Scotland College of
Agriculture, Craibstone
Farm, Bucksburn AB2 9TQ
Tel: 022-471 2677

Stock—chicks and pullets
**The following firms supply chicks and point-of-lay
pullets.**

Animal Medics Ltd (see
under Help and Advice)

Babcock Farms Ltd,
315–349 Mill Rd,
Cambridge
Tel: Cambridge 42031/4

Cyril Bason,
Stokesay, Craven Arms,
Salop.
Tel: Craven Arms 2304

Blue Barnes Hatchery,
West Acres, Pennyfine Road,
Sunniside,
Newcastle-on-Tyne
Tel: Whickham 887171

Denis E. Bonnett,
Longcroft Farm,
Ramsden Heath, Essex
Tel: Basildon 710197

William Brown,
Ballymacarn, Ballynahinch,
Co. Down, N. Ireland
Tel: Ballynahinch 2750

Castyne Ltd,
Semer House,
Semer,
Ipswich IP7 6JE
Tel: 0449 740975

Co-ordinated Hatchery Ltd,
Great Henny, Sudbury,
Suffolk
Tel: Twinstead 379

Stan Davies (Chicks),
The Rustings,
Poplars Drive, Marldon,
Paignton, Devon
Tel: Paignton 559789

Dekalb (UK) Ltd,
Refuge House, 3 Kings
Court, York
Tel: York 29137

Double-A Farms,
Bentham Lane,
Witcombe, Glos.
Tel: 045-282 3355

Elmbank Hatcheries,
Cavan, Irish Republic
Tel: Cavan 114

Greatock Poultry,
Tockington,
Bristol,
Avon BS12 4LQ
Tel: Almondsbury 613252

Green Hammerton
Hatcheries Ltd,
Hall Close,
Green Hammerton, York.
Tel: Green Hammerton
30384

Hamers Chicks,
Bradshaw, Bolton, Lancs.
Tel: Turton 852555

John Harpur,
Byeways,
Groton Street,
Boxford,
Colchester, Essex
Tel: Boxford 210543

M. J. Hayward,
Green Farm, North Gorley,
Fordingbridge, Hants.
Tel: Fordingbridge 52007

Ewart Hebditch Ltd,
South Somerset Poultry
Breeders,
Martock, Somerset
Tel: Martock 2599

Fred Horner (Hatcheries)
Ltd,
Dunnington, York, N. Yorks.
Tel: York 489 254

Hy-Line International,
Park Lane,
Mytholmroyd, Halifax,
W. Yorks.
Tel: Calder Valley 301

B. J. Ingram Ltd,
The Drift, Fakenham,
Norfolk
Tel: Fakenham 2755

H. M. Kirk,
Brewood Park Farm,
Coven, Wolverhampton,
W. Midlands
Tel: Wolverhampton 850284

Laurel Farm Eggs,
Laurel Farm,
Lower Claverham,
Bristol, Somerset
Tel: Yatton 833116

Hamish Morison Ltd,
West Morriston,
Earlston, Berwickshire,
Scotland
Tel: Earlston 250

Moy Park Ltd,
Donaghmore, Dungannon,
Co. Tyrone, N. Ireland
Tel: Donaghmore 362

Muirfield Hatchery,
Fossoway, Kinross, Scotland
Tel: Fossway 401

L. V. and B. Piggott,
68–70 High Street,
Eaton Bray,
Dunstable, Beds.
Tel: Eaton Bray 220944

G. Potter and Sons,
Howefield, Baldersby,
Thirsk, Yorks.
Tel: Melmerby 213

Quantock Hatcheries Ltd,
Pelham House,
Winchester Road, Goodworth
Clatford, Andover, Hants.
Tel: Andover 3157

Rhode Island Reds
Unlimited,
Broadway House,
Emborough, Bath,
Somerset
Tel: 0761 232286

A. E. Rodda and Son,
Turkey Farm,
Scorrier, Redruth,
Cornwall
Tel: St Day 820528

Ross Poultry,
Imperial House,
61–65 Rose Lane,
Norwich, Norfolk
Tel: Norwich 611161

Shaver Poultry Breeding
Farms (GB) Ltd,
Bawdeswell, Dereham,
Norfolk
Tel: Bawdeswell 254

South Western Chicks
(Warren) Ltd,
Broadway, Ilminster,
Somerset
Tel: Ilminster 3441

Southdown Hatcheries Ltd,
Uckfield, Sussex
Tel: Uckfield 4288

Southern Pullet Rearers,
Hawthorns,
8 Fishbourne Road,
Chichester, Sussex
Tel: Chichester 88551 and
Storrington 2681

Sunnyside Poultry,
Sunnyside Poultry Farm,
Little Green,
Bunwell, Norwich
Tel: Bunwell 507

Robert Thompson Chicks
(Warren) Ltd,
Lannhall, Tynron, Thornhill,
Dumfriesshire, Scotland.
Tel: Moniaive 329

John R. Todd Ltd,
Sidehead, Stewarton,
Kilmarnock,
Ayrshire, Scotland
Tel: Dunlop 244

Tracegrange Agricultural
Marketing Co Ltd,
The Poultry Farm,
Chearsley, Aylesbury,
Bucks.
Tel: Long Crendon 208335

Uglow's Accredited Farm,
Kingston, Stoke Climsland,
Callington, Cornwall
Tel: Stoke Climsland 206

Wessex Hatcheries Ltd,
Zeals, Warminster, Wilts.
Tel: Bourton (Dorset) 561

West Cumberland
Farmers Ltd,
Whitehaven, Cumbria
Tel: Whitehaven 3191

Whitakers Hatcheries Ltd,
Cauden Quay, Cork, Eire
Tel: Cork 53366

J. P. Wood and Sons
(Hatchery) Ltd,
Affcot, Marshbrook,
Church Stretton, Salop.
Tel: Marshbrook 218

R. A. Wright Chicks Ltd,
Portaferry Road,
Newtownards,
N. Ireland
Tel: Newtownards 3456

The following firms supply point-of-lay pullets only.

Abbot Bros (Est. 1876),
Thuxton, Norwich, Norfolk
Tel: Mattishall 220

W. C. Blacklocks Ltd,
Whitehall Farm, Lydd, Kent
Tel: Lydd 20593

Basil Carver,
Black Ness, E. Cornworthy,
Totnes, Devon
Tel: Dittisham 363

L. Chippindale,
Kingsley Poultry Farm,
Kingsley Road, Starbeck,
Harrogate, N. Yorks.
Tel: Harrogate 884042

Arthur Day (Corley) Ltd,
Meadow Rise Farm,
Church Lane, Corley,
Coventry, Warwicks.
Tel: Fillongley 40501

John S. Dunne,
Dunton Poultry Farm,
Bulphan,
Upminster, Essex
Tel: Basildon 43522

R. A. C. Heal,
Butlers Bank, Shawbury,
Shrewsbury, Salop.
Tel: 093 94 359

David Hitchings
(Broadchalke) Ltd,
Knapp Farm, Broadchalke,
Salisbury, Wilts.
Tel: Broadchalke 322

W. Potter and Sons
(Poultry) Ltd,
Fillongley, Coventry,
Warwicks.
Tel: Fillongley 40960

J. Rainford and Son Ltd,
Woodland Grange,
Penwortham,
Preston, Lancs.
Tel: Preston 44412

Ryder Poultry Enterprises,
Steep Marsh Farm,
Petersfield, Hants.
Tel: Liss 2235

S. R. Wheatcroft and Son,
Dover's Orchards, Hoo Lane,
Chipping Campden, Glos.
Tel: 0386-840366

The following firms supply chicks only.

Tom Barron
Hatcheries Ltd,
Catforth, Preston,
Lancs PR4 0HQ
Tel: Catforth 69011

David Chandler,
Lodge Farm,
Old Dalby,
Melton Mowbray,
Leics.
Tel: Melton Mowbray
822565

Chorley Chicks Ltd,
213 Preston Road,
Whittle le Woods,
Chorley, Lancs.
Tel: Chorley 2470

Churchstoke Hatchery Ltd,
Churchstoke, Powys,
Wales
Tel: Churchstoke 258

Euribrid Ltd,
Orchard House,
50–58 Pensby Road,
Heswell, Merseyside
L60 7RE
Tel: Chester 313325

H and N Inc,
Countess Road,
Dunbar,
East Lothian,
Scotland
Tel: 0368 62801

P. D. Hook (Hatcheries) Ltd,
Cote, Aston, Oxford, Oxon.
Tel: Bampton Castle 261

Horton Poultry,
Marford, Wrexham, Clwyd,
Wales
Tel: 024-459 577

Hubbard Poultry UK Ltd,
15 Gloucester Street,
Stroud, Glos.
Tel: 04536-4933

ISA Poultry Services Ltd,
Old Hall Hatchery,
Orton, Longueville,
Peterborough,
Cambs.
Tel: Peterborough 231131

A. E. Jennings,
Iron Cross,
Salford Priors,
Evesham,
Worcs. WR11 5SH
Tel: Evesham 870 321

James Rutter Ltd,
Fulletby, Horncastle, Lincs.
Tel: Tetford 666

Thornber Chicks Ltd,
Brier Hey, Mytholmroyd,
Hebden Bridge, W. Yorks.
Tel: Calder Valley 2641

Welford Hatcheries,
Boulderdyke Farm,
Clifton, Deddington,
Oxford
Tel: Deddington 230/484

Equipment suppliers

Brooders

Calor Gas Ltd,
Calor House, Windsor Road,
Slough, Bucks.
Tel: Slough 23824

Cope and Cope Ltd,
57 Vastern Road,
Reading, Berks.
Tel: Reading 54491

George Elt Ltd,
Eltex Works, Worcester
Tel: Worcester 422377

SBM (UK) Ltd,
13 David Road,
Poyle Trading Estate,
Colnbrook, Bucks.
Tel: Colnbrook 5353

Western Incubators Ltd,
Pauls and Whites
International (GB) Ltd,
Springfield Road,
Burnham-on-Crouch, Essex
Tel: Maldon 782999

Cages

Betterway Cages,
Bournemouth (Hurn) Airport,
Christchurch, Hants.
Tel: North Bourne 3898

Cope and Cope Ltd
(see Brooders)

Eggmen Ltd,
Victoria Industrial Estate,
Chorley Road,
Standish,
Wigan, Lancs.
Tel: Standish 424860

Home Egg Units,
Penygroes Farm,
Castellau, Llantrisant,
Mid Glamorgan
Tel: Llantrisant 350

Q-Mark International,
Hud Hey Road,
Haslingden, Lancs.
Tel: Rossendale 5323

Thornber Construction,
Hoo Hole Works,
Cragg Road, Mytholmroyd,
Hebden Bridge, W. Yorks
Tel: Calder Valley 2032

Drinkers and feeding equipment

Broiler Equipment Co Ltd,
Winnall, Winchester, Hants.
Tel: Winchester 61701

EB Equipment Ltd,
Redbrook, Barnsley, S. Yorks.
Tel: Barnsley 6896

George Elt (see Brooders)

South and Western
(Agriculture) Ltd,
Pen Mill Trading Estate,
Yeovil, Somerset
Tel: Yeovil 4915

Southern Pullet Rearers
(see Stock)

George Wilkinson
(Burnley) Ltd,
Progress Works,
Burnley, Lancs.
Tel: Burnley 26461

Egg cartons
Hartmann Fibre Ltd,
Kirk House, Birkheads Road,
Reigate, Surrey
Tel: Reigate 49241

Scan Packaging,
16 Earsham Street,
Bungay, Suffolk
Tel: Bungay 2979

Merchandising Aids,
67–69 Chancery Lane,
London WC2A 1AF
Tel: 01-405 4711

South and Western
(Agriculture) Ltd
(see Drinkers
and feeding equipment).

Houses
Domestic Fowl Trust,
Dorsington Manor,
Stratford-on-Avon, Warwicks.
Tel: Bideford on Avon 2442

Park Lines and Co,
501 Green Lanes,
London N13 4BS
Tel: 01-886 0011

Harry Hebditch Ltd,
Timber Building
Manufacturers,
Martock, Somerset
Tel: Martock 3883

R. J. Patchett,
Ryefield Works, Queensbury,
Bradford, W. Yorks.
Tel: Queensbury 882331

Hyline Rabbits Ltd,
Marston, Northwich, Cheshire
Tel: Northwich 41334

Southern Pullet Rearers
(see Stock)

A. E. Jennings
(see chick suppliers)

Sussex Joinery,
Adur Works,
Tanyard Lane,
Steyning, Sussex
Tel: Steyning 814135

Lewes Road Sawmills Ltd,
Glencoe, Scaynes Hill, Sussex
Tel: Scaynes Hill 451

Robert Miller (Denny) Ltd,
Scottish Appliance Works,
Broomhill, Bonnybridge,
Stirlingshire, Scotland
Tel: Bonnybridge 2424

Sydenham Hannaford Ltd
(C.A.),
Hamworthy, Junction,
Poole, Dorset BH16 5BP
Tel: 020 122 2363

Appendix B

Suggestions for further reading

The following Government publications are prepared by the Ministry of Agriculture. They can be ordered through local booksellers or by post from: PO Box 569, London SE1 9NH.

Priced publications
Control of Rats and Mice Bulletin 181, £1.50p.
Incubation and Hatchery Practice Bulletin 148, £1.75.
Intensive Poultry Management for Egg Production Bulletin 152, £1.75.
Poultry Housing and Environment Bulletin 212, £1.12p.
Poultry Nutrition Bulletin 174, £2.25.
The Small Commercial Poultry Flock Bulletin 198, 70p.

Free publications (Advisory Leaflets)
Artificial Incubation in Small Incubators A.L. 341.
Cannibalism and Feather Pecking in Poultry A.L. 480.
Culling Laying Hens A.L. 513.
Disinfection and Disinfestation of Poultry Houses A.L. 514.
Feeding the Modern Layer A.L. 446.
Feeding Pullet Replacement Stock A.L. 431.
Flies and Other Insects in Poultry Houses A.L. 537.
Free Range and Semi-intensive Systems for Egg Production A.L. 342.
Guide to the Choice of Laying Stock S.T.L. 130.
Heating of Laying Houses S.T.L. 102.
Keep your own Poultry Records S.T.L. 56.
Lighting for Egg Production A.L. 540.
Natural Hatching A.L. 43.
Preparation and Trussing of Poultry for Market A.L. 428.
Real Controlled Environment for Poultry S.R.L. 89.
Rearing Replacement Pullets A.L. 396.
Welfare Code No. 3 : Domestic Fowls.

Appendix C

Problems, their causes and cures

A. Egg problems	Possible causes and cures
(a) Lot of soft shelled eggs.	Normal if birds are coming into lay or nearing the end of lay. If it persists suspect calcium shortage and feed limestone grit. (See Chapter 7.)
(b) Mis-shaped eggs.	Not serious and eggs can be eaten; may clear up.
(c) Dirty eggs in nests.	Check that nest litter is not soiled—change regularly. Collect eggs regularly—at least 3 times a day. (See Chapter 7.)
(d) Blood on shell.	Normal in early lay providing not excessive i.e. only smears. Should clear up in a few weeks. If excessive or persistent, suspect injury, locate the bird and cull it. (See Chapter 7.)
(e) Blood or meat spots in the whites or in the yolk.	Not unusual and will disappear and re-appear from time to time. Nothing you can do.
(f) Very pale yolks.	Lack of pigment in the feed, Make fresh cabbage leaves available. (See Chapter 7.)
(g) Fewer eggs than expected.	Lots of possible reasons. Check for:

    (a) Pilfering.
    (b) Dog eating eggs.
    (c) Chickens eating own eggs—collect frequently.
    (d) Early moult—naked birds and lots of feathers on floor. (See Chapter 5.)
    (e) Mites, worms or coccidiosis. (See Chapter 6.)
    (f) Respiratory disease —listen to breathing and look for watery nose and eyes. (See Chapter 6.)

(h) Excessive number of cracked eggs.

(g) Water and feed availability and consumption. Also sour feed and dirty water. (See Chapter 6.)
(h) Feather pecking.
(i) Lighting—have you compensated for decreasing daylight after June 21? (See Chapter 5.)
(j) Broodiness—isolate offenders in broody coop. (See Chapter 5.)
(k) Very hot or very cold weather.
(a) Check method and frequency of collection.
(b) Shortage of calcium—feed limestone grit.

B. Bird problems

(a) Excessive loss of feathers part-way through lay.

Early moult—caused by sudden stress like food and water shortage or lighting problems. Put right.

(b) Feather pecking.

Boredom, feed shortage or damage to one bird that attracts others. Sort out cause and treat injuries with anti-peck ointment. (See Chapter 6.)

(c) General loss of condition coupled possibly with few eggs, lack of appetite, weight loss, birds standing around with feathers ruffled. Possibly runny droppings white, light brown or grey in colour.

Could be many things but if only one or few birds affected, isolate these from the rest to a warm dry place with plenty of food and water. Give worming powders. Also check for mites, lice and fleas, freshness and availability of feed and water. (See Chapter 6.)

(d) Droppings more moist than usual.

Possibly too much salt in the feed (likely if scraps are fed), sudden onset of enteritis or very hot weather which encourages excessive drinking.

(e) Some deaths and severe respiratory symptoms accompanied by fewer eggs.

Suspect fowl pest—inform the Ministry of Agriculture. (See Chapter 6.)

(f) Bird persists in sitting on eggs—pecks at hand if placed near. Stops laying.

Gone broody. Isolate in broody coop to break the habit. (See Chapter 5.)

(g) Swollen or distended crop area—hard or soft lump.

Crop binding or sour crop. Can be treated but if persists—cull. (See Chapter 6.)

(h) Part of the innards protrude from the vent, possibly when laying large egg.

Prolapse due to excessive straining. Can be treated but rarely lasting cure. Best to cull if persists. (See Chapter 6.)

(i) Bird frequently crouches as if to lay, but no egg. May wander round in a crouching position—egg may just be visible at vent.

Egg binding—may be due to particularly large egg. Can attempt to treat the bird but extreme care needed. (See Chapter 6.)

Appendix D

Daily, weekly and occasional tasks

Daily

(a) Collect eggs three times a day.
(b) Shut up birds at night and release in morning.
(c) Check lights and timing mechanisms, if any, are working properly. (See Chapter 5.)
(d) If feeding wet mash and scraps, ensure it is all eaten up and wash feed trough. If dry mash check plenty available.
(e) Make sure plenty of water available and wash drinkers.
(f) Make habit of observing the birds every day for early signs of problems. Watch the droppings and check that house litter is in good condition.

Weekly

(a) Clean droppings boards or droppings pit.
(b) Check for presence of mites. (See Chapter 6.)
(c) Where feeding dry mash, allow birds to empty feeder once a week to avoid stale feed. Re-fill.
(d) Ensure you have plenty of feed to carry through following week.
(e) Watch the condition of the run, in bad weather or if for some reason the run gets very muddy shut the birds up in the house continuously and treat the run. Do not release birds until run is fit. (See Chapter 2.)
(f) Look for evidence of mice, rats or foxes and take appropriate action if found. (See Chapter 6.)
(g) Sterilize drinkers in boiling water.

Occasional

(a) If mites are found, treat the house twice a year and give a coat of creosote. (See Chapter 6.)
(b) At end of laying year or other time when house depopulated, remove and dismantle all furniture and fittings, dig out litter and give thorough clean down as described in Chapter 6.

Appendix E

Lighting schedule

	hours of daylight	hours of artificial light for 17 hr day	Switch-on time for morning light
January			
1st week	8.00	9.00	00.05
2nd week	8.11	8.49	00.13
3rd week	8.27	8.33	00.24
4th week	8.47	8.13	00.36
February			
1st week	9.10	7.50	00.49
2nd week	9.35	7.25	01.02
3rd week	9.59	7.01	01.14
4th week	10.26	6.34	01.27
March			
1st week	10.54	6.06	01.40
2nd week	11.21	5.39	01.52
3rd week	11.49	5.11	02.04
4th week	12.17	4.43	02.16
5th week	12.43	4.17	02.27
April			
1st week	13.11	3.49	02.30
2nd week	13.39	3.21	02.51
3rd week	14.06	2.54	03.03
4th week	14.31	2.29	03.14
May			
1st week	14.56	2.04	03.25
2nd week	15.20	1.40	03.37
3rd week	15.42	1.18	03.48
4th week	16.01	0.59	03.58
June			
1st week	16.16	0.44	04.06
2nd week	16.28	0.32	04.13
3rd week	16.35	0.25	04.18
4th week	16.38	0.22	04.21
5th week	16.36	0.24	04.21
July			
1st week	16.28	0.32	04.18
2nd week	16.17	0.43	04.14
3rd week	16.03	0.57	04.07
4th week	15.44	1.16	03.58
August			
1st week	15.23	1.37	03.47
2nd week	15.00	2.00	03.35
3rd week	14.35	2.25	03.21
4th week	14.10	2.50	03.07
5th week	13.44	3.16	02.52

	hours of daylight	hours of artificial light for 17 hr day	Switch-on time for morning light
September			
1st week	13.16	3.44	02.36
2nd week	12.49	4.11	02.20
3rd week	12.22	4.38	02.04
4th week	11.55	5.05	01.48
October			
1st week	11.27	5.33	01.32
2nd week	11.01	5.59	01.17
3rd week	10.34	6.26	01.02
4th week	10.07	6.53	00.47
November			
1st week	9.41	7.19	00.34
2nd week	9.17	7.43	00.22
3rd week	8.55	8.05	00.12
4th week	8.34	8.26	00.03
5th week	8.16	8.44	23.56
December			
1st week	8.03	8.57	23.52
2nd week	7.54	9.06	23.51
3rd week	7.51	9.09	23.53
4th week	7.52	9.08	23.57

Footnote: The above schedule provides for a constant 17 hour day and indicates the time at which the lights should come on in the morning. The problem of giving the birds an opportunity to roost can be avoided by using this method of putting the lights on in the morning. Birds will then roost naturally as the light fades in the evening. There will be no chance of them being stranded on the floor when the lights suddenly go out. If you prefer a 14 or $14\frac{1}{2}$ hr day just subtract 3 or $2\frac{1}{2}$ hours from the amount of artificial light used and bring the switching-on times forward by the appropriate amount. Times given are Greenwich Mean Time so adjustments will have to be made for British Summer Time.

Index